Liquid Chromatography
of Polymers and
Related Materials

CHROMATOGRAPHIC SCIENCE

A Series of Monographs

Editor: JACK CAZES
Fairfield, Connecticut

Other Volumes in Preparation

Liquid Chromatography of Polymers and Related Materials III

edited by

JACK CAZES

Waters Associates, Inc.
Milford, Massachusetts

CRC Press
Taylor & Francis Group
Boca Raton London New York

CRC Press is an imprint of the
Taylor & Francis Group, an **informa** business

First published 1981 by Marcel Dekker, Inc.

Published 2019 by CRC Press
Taylor & Francis Group
6000 Broken Sound Parkway NW, Suite 300
Boca Raton, FL 33487-2742

© 1981 by Taylor & Francis Group, LLC
CRC Press is an imprint of Taylor & Francis Group, an Informa business

First issued in paperback 2019

No claim to original U.S. Government works

ISBN 13: 978-0-367-45200-1 (pbk)
ISBN 13: 978-0-8247-1514-4 (hbk)

**Visit the Taylor & Francis Web site at
http://www.taylorandfrancis.com**

**and the CRC Press Web site at
http://www.crcpress.com**

Library of Congress Cataloging in Publication Data
Main entry under title:

Liquid chromatography of polymers and related materials
    III.

    (Chromatographic science ; v. 19)
    Papers from International GPC Symposium/80: GPC/LC
Analysis of Polymers and Related Materials, held
Oct. 23-24, 1980, in Framingham, Mass.
    Includes indexes.
    1. Polymers and polymerization--Analysis--Congresses.
2. Liquid chromatography--Congresses. I. Cazes, Jack,
[date]. II. International GPC Symposium/80: GPC/LC
Analysis of Polymers and Related Materials (Framingham,
Mass.) III. Series.
QD139.P6L56   547.7'046   81-5434
ISBN 0-8247-1514-4   AACR2

PREFACE

Published in this volume are papers presented at the International
GPC Symposium/80: GPC/LC Analysis of Polymers and Related Materials,
which was held on October 23-24, 1980 at the Sheraton Tara Hotel in
Framingham, Massachusetts. This volume, the third of a series,*
describes new GPC/LC applications and techniques that will provide
polymer scientists and practitioners with insight into the development of new polymers and plastics and improvement of existing
materials.

I thank the authors of the contributed papers for the fine work
they have done and reported here, and also for their patience in the
preparation of their manuscripts.

Special thanks are extended to Mrs. Nancy Leutert for her valuable assistance in the preparation of the indexes for this volume.

Lastly, thanks to Waters Associates, Inc. for sponsoring the
symposium and for making their facilities available during the
preparation of the final manuscript.

Jack Cazes

---

*The first two volumes of this series are: Liquid Chromatography of
Polymers and Related Materials (J. Cazes, ed.), Marcel Dekker, New
York, 1977; and Liquid Chromatography of Polymers and Related
Materials II (J. Cazes and X. Delamare, eds.), Marcel Dekker, New
York, 1980.

CONTENTS

CONTRIBUTORS

MICHAEL R. AMBLER   The Goodyear Tire and Rubber Co., Akron, Ohio

C. D. CHOW   Analytical Laboratories, Dow Chemical U.S.A., Midland, Michigan

WILLIAM A. DARK   Waters Associates, Inc., Milford, Massachusetts

DEBORAH K. HADAD   Lockheed Missiles and Space Company, Inc., Sunnyvale, California

G. L. HAGNAUER   Polymer Research Division, Army Materials and Mechanics Research Center, Watertown, Massachusetts

W. HEITZ   Fachbereich Physikalische Chemie, Polymere, Philipps-University, D-3550 Marburg, Federal Republic of Germany

MOLLY Y. HELLMAN   Bell Laboratories, Murray Hill, New Jersey

R. M. HOLSWORTH   Glidden Coatings and Resins, Division of SCM Corporation, Strongsville, Ohio

G. E. JOHNSON   Bell Laboratories, Murray Hill, New Jersey

A. F. KAH   Glidden Coatings and Resins, Division of SCM Corporation, Strongsville, Ohio

T. N. KOULOURIS   Polymer Research Division, Army Materials and Mechanics Research Center, Watertown, Massachusetts

C. KUO   Glidden Coatings and Resins, Division of SCM Corporation, Strongsville, Ohio

M. W. LONG, JR.   Analytical Laboratories, Dow Chemical U.S.A., Midland, Michigan

BENJAMIN MONRABAL*   Chemistry Department, Virginia Polytechnic Institute, Blacksburg, Virginia

JOHN C. MOORE†   The Dow Chemical Co., Freeport, Texas

T. PROVDER   Glidden Coatings and Resins, Division of SCM Corporation, Strongsville, Ohio

J. G. ROONEY   Elastomers Technology Division, Exxon Chemical Company, Linden, New Jersey

LARRY E. STILLWAGON   Bell Laboratories, Murray Hill, New Jersey

GARY N. TAYLOR   Bell Laboratories, Murray Hill, New Jersey

G. VER STRATE   Elastomers Technology Division, Exxon Chemical Company, Linden, New Jersey

LOWELL WESTERMAN   Plastics Technology Division, Exxon Chemical Company, Baytown, Texas

---

*Current affiliation: Dow Chemical Iberia, Tarragona, Spain.
†Now retired.

GEL PERMEATION CHROMATOGRAPHY: A TWENTY YEAR VIEW

John C. Moore*

The Dow Chemical Co.
Freeport, Texas

ABSTRACT

Reviewing the earlier years of GPC, a few subjects seemed to be
unfinished business, and these are discussed. The inception of GPC
is cited as an example of the value of a strong driving force and a
diverse background in innovation. In support of Casassa's early
view of the separation mechanism in GPC, a detailed picture is
offered of the internal pore structure of Haller's porous glass. An
average pore diameter of 2400 A is found for a glass which showed
1800 by mercury penetration. This brings theory and experiment much
closer together. Also, the effects of diffusional non-equilibrium
are discussed in their contribution to zone-broadening rather than
to earlier elution.

I.   INTRODUCTION

For me this fall of 1980 is the twentieth birthday of gel permeation

chromatography--GPC. There are still a few things about it that to

me are unfinished business and so I would like to get some remarks

into the record. First I would like to comment a bit further on my

experience at the start of GPC, as an example of the value of a

strong driving force and a diverse background in innovation, and

---

*Now retired.

1

then I will go on to some aspects of the mechanism of the separation.
I do have a little data to present, but mostly I want to give you my
picture of what is going on in those little porous beads, for whatever it may be worth to you.

II. HISTORICAL

I have already told about the inception of GPC [1], but there is a
story behind that. In 1960 I had been looking closely at liquid
chromatography as a research area. Having been involved with
development in gas chromatography for several years I thought that
continuous operation with high performance columns and plumbing
should be made generally available in LC. At that time, much of LC
was geared to packing a fresh column for each sample to achieve
reproducible adsorption. Column efficiency was rarely noted but it
was low, around 150 plates per foot, and it got lower as column
diameter increased. The cause was uneven packing and the effect
was called channelling. Before GPC could come out that had to be
dealt with too. Then one day I ran across the report of Porath and
Flodin on Gel Filtration [2]. They had a short broad bed in a
Buchner funnel; proteins were excluded from crosslinked dextran
particles while salts were delayed by their admission to the gel
particles. This was not the first time that molecular size exclusion had come to my attention, but this time I was open for a new
project, and things came together. From long association with the
manufacture and uses of ion exchange resins, the GC and LC exposures,
and a deeply felt need for a good method of getting molecular weight
distributions, the vision of GPC took shape rather quickly.

  This deeply felt need seems to have been the crucial factor in
getting my attention on my opportunity. It arose in the early 1940's
when I was developing some polypropylene glycols. Ionically catalyzed, of relatively low molecular weight, their distribution was a
clean, narrow Poisson function but we couldn't prove it then. This
was painful because our salesmen were sending notes like "this customer was told by our competitor's man that our product doesn't have

a good molecular weight distribution." I was sure he was bluffing, but I didn't know how to handle it. The situation passed quickly but the remembrance did not. After this, anything on molecular weight distribution seemed to catch my attention. I believe this was the subconscious flag that waved when the GPC idea was ready to come forth. This was the source of energy that powered my initial exploration, these polyglycols were in fact my first samples, and my program to tailor-make a series of column packings with the complete range of permeabilities was started when I found the pores of the standard ion-exchange beads were all too small and those of the large-pore beads were too large for those polyglycols. It was also apparent then that we needed much smaller beads with more pore volume and more rigidity, and the means for these were available. Being aware, then, of the widespread availability of these ingredients in my vision of GPC, I thought it very possible that others might have had similar experiences and might even then be working along the same lines. We later found that we were probably no more than a year ahead of others in the basic idea, so my sense of urgency was justified. But it is pleasant to recall that in a year and a half from starting work to first public announcement [3] the basic details of the instrument and its utility were soundly established.

III. SEPARATION MECHANISM - EQUILIBRIUM

Now let us turn to the separation mechanism in gel permeation chromatography. In 1966 I presented some data on columns packed with a very uniform pore glass, kindly made available to me by Dr. Haller at the National Bureau of Standards [4]. The data showed that we were getting some penetration by pretty large molecules, compared to the pore size found by Haller with mercury intrusion, and I suggested that the whole range of conformational sizes of a macromolecule might be involved in its GPC "size" [5,6]. Haller had established that the rigid rods of tobacco mosaic virus were

excluded completely from glass pores much larger than the rod's
diameter, but much less than its length. So we had the picture
that the molecule's tumbling was faster than its translational
diffusion, and that was faster than the change in domain size of a
flexible macromolecule. Dr. Ed Casassa at Mellon Institute then
took the case of equilibrium in a theta solvent, and calculated the
pore penetration of a random coil molecule as that fraction of all
its conformations that would not touch the wall of the pore [7].
Dr. Turner Alfrey at Dow had also made such a calculation using a
different mathematical treatment but coming to the same conclusion
[8]. The pore penetration I reported fitted well with their calculations for a slab-shaped pore of the diameter found by mercury
intrusion, but was too great for a tubular pore of that diameter.
Please note here that the meaning of K_d, the volumetric distribution
coefficient, is here defined implicitly as the ratio of a solute's
concentration in the pore to that outside the pore. We will come
back to this later.

Now I will offer a correction to the pore diameter we cited
then. When a scanning electron microscope (SEM) became available
to us, I took the matter up again. We made a number of specimens
by potting several grains of Haller's 1800 Angstrom pore glass in
gelatin capsules with styrene containing 25 percent divinylbenzene.
After opening with a microtome, we used reagent hydrofluoric acid
to eat away the glass and expose the pore structure. After washing,
mounting and gold-flashing they were examined under the SEM. First
we found that an hour or two of etching was nowhere near enough, it
took twenty-four hours to stabilize and clear the picture. Then at
30,000 times magnification in an 8 x 10 inch print we had a remark-
ably clear picture of the pore structure (Fig. 1). Since the pores
were formed by the nucleation and growth of a soluble phase in a
plastic mass, with coalescing of nuclei, we should expect a recurrent pore diameter, with enlargements, branchings and joinings. I
picked out on the print over a hundred of these rather uniform necks
at random, measured them, ranked and plotted the measurements at
five percent intervals on a linear vs probability basis, shown in
Figure 2. They fitted well on a straight line, with mean neck

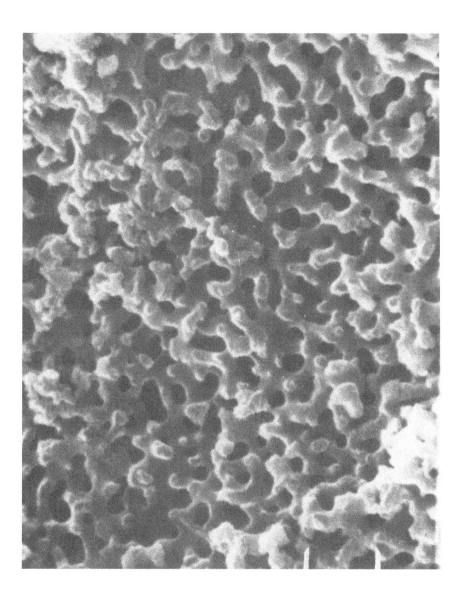

FIGURE 1. Scanning electron micrograph of the pore system of
Haller's 1800 Angstrom pore glass, potted in styrene + 25% divinylbenzene, with glass removed.   Between points = 1 micrometer.

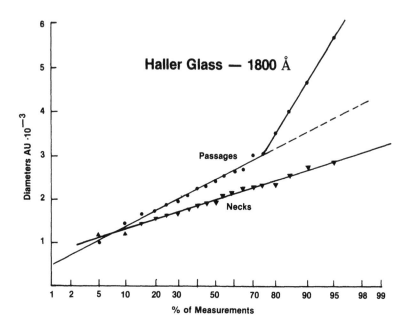

FIGURE 2. Pore diameters in thousands of Angstroms vs percent of measurements, from specimen as above.

diameter of 1970 Å, standard deviation 530 Å. From this picture we might expect a very uniform mercury intrusion pressure corresponding to these recurrent neckings in the pore structure. Since the gold flashing could be 100 Å on top, less on the sides were measured, and both the SEM and the mercury penetration were subject to calibration, we felt this to be a gratifying agreement. I also marked and measured 120 pore widths at random, including enlargements, and these gave a mean pore diameter of 2440 Å, 24 percent greater. The slope also was greater, with standard deviation 850 Å up to 75 percent of the readings, and above that the plot bent sharply upward to 6000 Å diameter at 96 percent, due to multiple intersections. So with this picture I consider it likely that Casassa's and Alfrey's concept of the separation mechanism is in good agreement with my data.

In GPC theories the straight line calibration curve observed with the lower permeability gels is usually cited as the normal and proper thing. Although we all love to find nice straight lines, it is awkward that these have to become vertical at both ends. Let us

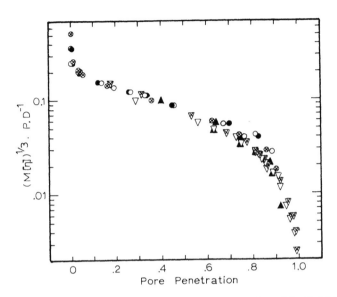

FIGURE 3. Calibration curves, pore penetration $(V_e-V_o)/(V_t-V_o)$ vs. cube root of hydrodynamic volume (M.W. x int. vis.) divided by pore diameters from mercury intrusion. Circles = 242 A, Triangles = 1800 A. Solids = untreated silica glass, polar theta solvent (6 2-butanone + 1 methanol). Open = polar theta solvent, siliconized pores, X in symbol = tetrahydrofuran, siliconized surfaces.

note here that the gels have a more or less random distribution of internal structures, while our data on Haller's nearly uniform pore glass, in Figure 3, showed a calibration curve much more like an integral probability curve, although steepened a bit by the slight diversity in pore structure and altered a bit by the logarithmic scale of the molecular size. If we were to mix equal amounts of ten of these porous glasses, with their exclusion points going up by a factor of two to five in each successive one, the resulting calibration would indeed have a long straight section and it would also become vertical at both ends. With a single gel containing the proper randomness in the distribution of its pore sizes we should have no difficulty in explaining an extended straight section in the calibration curve. Our composite columns often do give nearly straight calibration lines over a very extended range. This should be attributed to a skillful or fortunate choice of gel structures, and not to the fulfillment of some theoretical condition.

Even if it be granted that we have good agreement between equilibrium theory and peak locations at moderate flow rates, there

are still many questions. Some of these seem to me to have false
premises underlying them. For example, how can pores so uniform
show such an extended separation range? Or, how can molecules
smaller than the pore show any separation except by their varying
diffusion rates? Then of course the third follows: if a slower
diffusion rate can cause earlier elution, then why are the elution
peak positions so nearly independent of flow rate? For the first
question, remember that the pore has no attractive force with which
to stuff itself full of molecules. Rather, K_d = 1 means that the
solute concentration in the pore can rise to the level outside.
Now since the large rod-shaped virus behaved as if it were spinning,
and the coil polymer molecule is rotation also, if we consider a
spherical molecule we have a fairly general case. Let us postulate
a sharp circular entrance to the pore, and that a molecule can only
enter by a direct hit. Where it may strike with its center anywhere
within the pore circle, it cannot enter unless its center comes within a smaller circle which is one molecule radius away from the wall
and concentric with it. Its probability of entrance, then, is the
ratio of the areas of these two circles. For a molecule with diameter equal to the pore the smaller circle is only a point, with no
area, so K_d = 0. A molecule with half that diameter will have a
pore concentration one-fourth that outside, $K_d = (1 - .5)^2$ or 0.25,
if one-tenth then $K_d = (1 - .1)^2 = 0.81$. So even without the coil
molecule's range of conformational sizes, just the probability of
admission to the pore gives a considerable separation range.

It is generally agreed that GPC separates on the basis of
molecular size. Can we agree now that GPC should be giving us an
absolute measure of molecular size? We should not think of it as
only an empirical method, that unexplainably seems to give equilibrium results under non-equilibrium conditions. I believe we have
here a well-defined pore system and a useful picture of the separation process, and a tentative figure can be obtained from this. In
the work I reported in 1966, and confirmed in subsequent work with
two pore sizes of Haller's glass, we found very close agreement

between their calibration curves when we plotted relative pore
penetration, K_d, against the ratio of hydrodynamic diameter to the
pore diameter, as shown in Figure 3. At 50 percent penetration,
by Casassa's and Alfrey's reasoning, the pore diameter should
represent the average conformational diameter. At this point, using the mercury intrusion diameters, the ratio of the cube root of
hydrodynamic volume to pore diameter was 0.075; with pore diameter
increased 35 percent to agree with our SEM data the ratio was
0.0555. At half of the reciprocals of these numbers, the average
solvated radius should lie between 6.8 and 9 times the cube root of
the hydrodynamic volume (molecular weight × intrinsic viscosity) or
about 2.5 × the RMS radius of gyration. So, when you use any column
with the hydrodynamic volume calibration, haven't you really obtained
this molecular size information for every chromatogram?

IV. DIFFUSIONAL NON-EQUILIBRIUM

The second and third questions deal with diffusional non-equilibrium.
A separation mechanism based on restricted diffusion appeared very
early in GPC and has been reviewed elsewhere [9,10]. With these
corrected sizes for the glass pores I feel that the need for it has
disappeared, if indeed there was any. Diffusion rates are important,
as we shall see, but diffusion distances are important also. Even
in the lower permeability Styragels, electron micrographs showed
that the high degree of crosslinking in the presence of a diluent
gives a reticular structure rather than a classical random network
[11]. The solid part is very tight, and the pore walls are probably
clothed with wasted loops and ends of chains, but the pores are
open. As the micrographs show, in Styragel the larger the pore the
less likely it is to extend far at that size, so that few large
openings will touch the bead surface. Therefore with Styragel the
larger molecules must diffuse out the way they got in, and they
don't have far to go either way. Porous silica and glass have even
more sharply defined pore walls, and they have their own structural

details. With structures like the Haller glass the larger molecules do not have shallower pores but there may be a very slow flow
through the particles.

With that view of bead structure, let us look at flow rate.
We know that the narrowest peaks come with small samples of very
fast-diffusing materials, at very slow flow rates, and of course
with the best columns and low-viscosity solvents. We know that
totally-excluded samples of polymers are broader in peak width even
though they come out earlier, so we can agree that diffusion rate
is important in the moving phase too. The broadest peaks come with
molecules that diffuse slowly and do spend time in the pore system,
and of course faster flow rates broaden all peaks. Now let us
examine the concentrations inside the pores. At low flow rates with
very small molecules we may expect that the concentration in the pore
will be substantially in equilibrium with that outside the particle;
the concentration in the pore is substantially level, rising and
falling with that outside the particle as the peak passes by. At
faster flow rates and with larger molecules there can be a concentration gradient within the pore. Let us look closely at that gradient.
As long as the external concentration is rising at an increasing
rate the pore concentration gradient will be rising also. Relatively less of the solute will be getting into the pore, and more will
be passing by because of this gradient, and so the eluted peak will
start up earlier. The peak concentration may even pass by the pore
opening while the depth of the pore still has zero concentration of
the solute. As the solute concentration falls outside, the pore
concentration starts to fall near the entrance, and a reverse gradient starts near the mouth of the pore. The pore will then contain
a concentration peak of its own! Now note that this peak within the
pore is being lowered by diffusion farther in at the same time that
diffusion out is occurring. Even with zero concentration outside,
there is no way for the depth of that pore to remain unexplored.
No way! Therefore, just as non-equilibrium makes the peak broader
on its rising side it must also make it broader on its falling side.

The concentration in the depth of the pore must continue to rise until what is left of the peak within the pore has gone all the way in, and then diffusion out will prevail throughout the pore. With this picture we can see why the peak width of a very narrow polymer sample increases as molecular weight increases, and then decreases again near the exclusion limit as available pores become shallower and finally disappear, leaving us with only the interstitial flow broadening effect. And I think we can see why the peak position changes so little with flow rate.

V. CONCLUSION

I am satisfied, then, that separation in GPC is based simply on steric exclusion. If we make a successful correction for zone broadening, then we have the effect of an infinite plate column and we have also corrected to diffusional equilibrium. One further correction may be needed. The hydrodynamic volume of a molecule, being based on its intrinsic viscosity, is expressed in terms of zero concentration. At usable concentrations in the column the macromolecular domains do interact by compressing each other. The correction of apparent molecular size to zero-concentration size may need to be incorporated in the zone-broadening correction. Some work has appeared along this line, and I plan to offer some data on it in another paper.

REFERENCES

1. J. C. Moore, J. Polym. Sci., C21, 1 (1968).

2. J. Porath and P. Flodin, Nature, 183, 1657 (1959).

3. Gordon Conference on Separations, New London, N. H., August 1962.

4. W. J. Haller, Chem. Phys., 42, 686 (1965).

5. J. C. Moore and M. C. Arrington, Preprints, Third International Seminar on GPC, Geneva, May 1966.

6.   J. C. Moore and M. C. Arrington, Preprint VI-107, International Symposium on Macromolecular Chem., Tokyo-Kyoto, 1966.

7.   E. F. Casassa, J. Polym. Sci., B5, 773 (1967).

8.   T. Alfrey, Jr., private communication.

9.   J. F. Johnson and R. S. Porter, Progr. Polym. Sci., 2, 203 (esp 210), 1970.

10. Fractionation of Synthetic Polymers, L. H. Tung ed., Marcel Dekker, New York, 1977, pp. 566-568.

11. Polymer Fractionation, M. J. R. Cantow ed., Academic Press, New York, Figure 4, p. 149.

MACROMOLECULAR COMPRESSION AND VISCOUS FINGERING
AS DEMONSTRATED IN FRONTAL GPC

John C. Moore*

The Dow Chemical Co.
Freeport, Texas

ABSTRACT

Studies of rerun cuts and of chromatograms with short, efficient
columns and with sample mixtures of narrow polystyrenes showed that
stirring and distorting effects were present in the peaks at all but
very low concentrations. The overload effect called viscous fingering was identified in these data, and experiments with frontal analysis were undertaken to study GPC in the absence of viscous fingering.
Another overload effect then became evident, a delay of the peak area
due to mutual compression of the macromolecular domains. Equations
from work by S. H. Maron and R. B. Reznick were adapted for correction of this effect, both in frontal and in small sample chromatograms.

I.   INTRODUCTION

Efforts to understand and improve GPC have been going on since its

earliest days. The time of analysis was a focal point for many of

these efforts. Where, in other types of liquid chromatography the

usual separation volumes run between two and five column volumes,

here the separation must appear between about half and one column

volume. Since the separation factors are thus limited the total

---

*Now retired.

number of plates must be much greater, and columns of twelve to
twenty feet were commonly used. While smaller and more uniform beads
were being developed, and packing methods were being improved, it was
also recognized that computer correction of the zone broadening effect should be a way to complete a separation that was only begun in
the column, and then shorter columns could be used. Both flow re-
versal and rerun cut techniques were used in this study, and the
latter showed that there was a stirring effect under the main peak,
even at quite moderate loads. The load effect called viscous fingering was identified in this, and reported in 1970 [2]. Since
viscous fingering is caused by a lower viscosity fluid pushing un-
evenly into a higher viscosity zone it is most important with short
columns of high efficiency, just the case we wanted to study. Rather
than be limited by this to the very smallest samples at the highest
detector sensitivities, already a well explored situation, we decided to explore frontal analysis, in which viscous fingering effects are delayed until after the analysis.

II. HISTORICAL

Reference to frontal analysis has been rare in the literature of
GPC. Winsor and co-workers used it to study the column zone broadening effect [3,4] and to detect interaction between macromolecules
[5]. Its classical application to chromatography under strong adsorption conditions was shown with polystyrenes in carbon tetrachloride on macroporous silica gels by Kiselev [6]. Although the
frontal technique was not used, an elution delay was observed by
K. Hellsing [7] for albumin when dextran or other neutral polymers
were present in the aqueous eluent. He observed a shrinking effect
of the solution on the Sephadex G-200 gel but did not postulate any
for the albumin, ascribing the delay to a solubility increase. The
concept of macromolecules shrinking each other's domains, rather
than interpenetrating, was clearly formulated by Maron and his
students some time ago [8]. Working with a highly developed

viscosimeter, he noted a straight-line relation between polymer concentration and the reciprocal of its "effective volume factor." Reznick [9] elaborated on this, finding it valid in good solvents, and in poor solvents except at high molecular weights where slight curvature appeared at very low concentrations [10]. Here we have made use of Reznick's version.

III. EXPERIMENTAL

For frontal analysis our breadboard GPC was altered by adding a 30 ml sample loop which was a vertical helix of quarter inch tubing with an inch or so turned up at the top and down at the bottom, and connected to the sample valve with our usual 0.020 inch I.D. transfer tubing. Also the Waters Differential Refractometer with 10 microliter cell volumes was fitted with a delay loop, coiled up inside the cell box, through which the sample and reference cells were put in series. Thus the recorder would show the first derivative of the frontal chromatogram. Delay volumes of 0.25 to 0.67 ml worked well, with about 0.5 ml used for a 30 × 3/8 inch column. Eluate volumes were measured with a photoelectrically operated dual pipette system. The 10-mv recorder already had wide range attenuator and zero suppression circuits, of Dow Chemical USA design. Columns ranged from 1 foot × 3/8 inch OD to 4 feet × 3/8 inch OD, generally packed with mixed gels closely graded, under 20 micrometers in diameter.

IV. RESULTS AND DISCUSSION

It seemed desirable to document the phenomena we encountered in getting acquainted with frontal analysis in GPC. Figure 1 shows two chromatograms run on a column 3/8 inch × two feet, with a 0.25 ml differentiating loop in the refractometer cell box. The volume measuring pipettes were set at one milliliter. The sample was a standard narrow polystyrene of M_w = 160,000, made up to 2 mg/ml of

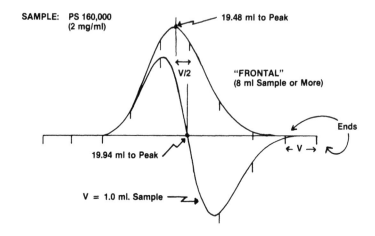

FIGURE 1. Frontal sample 8 ml or more, below: small sample 1.00 ml Column 2 ft × 3/8 inch, 0.25 ml differentiating loop. Samples: PS M_w 160,000, 2 mg/ml in tetrahydrofuran.

tetrahydrofuran. Run at 1 ml per minute, a sample of one ml volume shows the first derivative of its usual shape, starting up about 19 ml, giving a positive peak at the usual first inflection point, then steeply crossing the baseline at 19.94 ml for the peak location, then a negative peak for the second inflection point and ending about 24 ml. The frontal curve, from a sample 8 ml or more in volume, shows a conventional shape. It starts up at the same volume, but peaks earlier, at 19.48 ml. Since it developed from the start of the sample flow instead of from the center of the small sample, it should be 0.50 ml earlier and it was only 0.04 ml later than that. Also it ends about 1 ml earlier, so delay ascribable to polymer concentration is very small here.

Figure 2 compares the same frontal chromatogram, starting with pure solvent and ending with the sample concentration, with the flushing chromatogram. This is not a vacancy chromatogram from a 1 ml sample of pure THF, but the first derivative of the fall of concentration from 2 mg/ml to zero. The only significant load effect showing here is the later start-up of the flushing chromatogram,

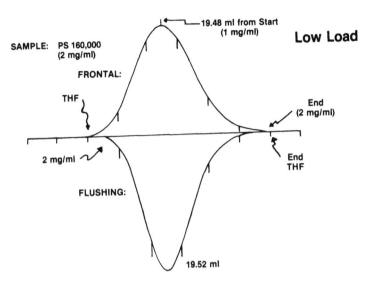

FIGURE 2.   Above: Frontal chromatogram from start of sample flow.
Below: Flushing chromatogram from end of sample flow.   Sample and
equipment as in Figure 1.

High Load

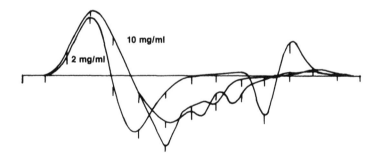

FIGURE 3.   Molecular compression and viscous fingering in small
samples, with differentiating loop in refractometer.   Polymer,
column and solvent as in Figure 1, 1.00 ml samples.   Below, 2 mg/ml.
Above, 10 mg/ml twice.

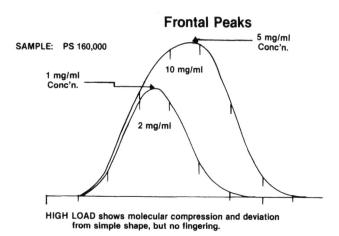

FIGURE 4. Frontal analysis, system as in Figure 1. Below: 2 mg/ml, above: 10 mg/ml shows molecular compression, deviation from simple shape but no fingering.

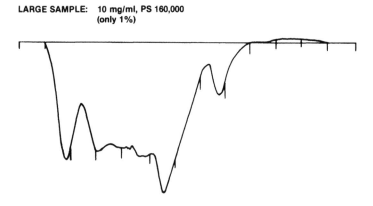

FIGURE 5. Flushing chromatogram from 10 mg/ml sample in Figure 4 shows fingering.

Frontal Chromatograms

1) 0.25 mg/ml PS 2 x 10^6 MW
2) 0.25 mg/ml PS 4 x 10^5 MW
3) 0.25 mg/ml of each

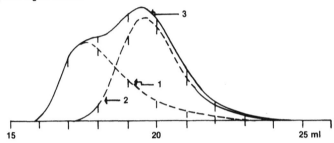

FIGURE 6. Frontal analysis of mixtures: system as in Figure 1, polystyrenes 2.15×10^6 and 4.11×10^5 at 0.25 mg/ml, separately and in mixture.

Frontal Chromatograms

1) 0.25 mg/ml PS 2 x 10^6 MW
2) 0.25 mg/ml PS 4 x 10^5 MW
3) 1.0 mg/ml of each

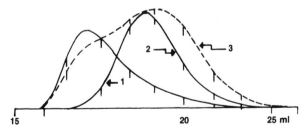

FIGURE 7. Frontal analysis showing compression effect, samples as in Figure 6, dashed line at 1.00 mg/ml each.

saying that the solute molecules are that much smaller at close to
sample concentration. This contrasts with the situation in Figure
3. Here the 2 mg/ml small sample shown in Figure 1 is repeated, with
two other 1 ml samples, same polymer but 10 mg/ml, or 1 percent. We
see that the higher concentration delays the first inflection point,
shown as a positive peak, only a little. The peak concentration,
shown as the baseline crossing, comes about 0.8 ml later because of
the molecular compression effect. The curves are smooth up to this
point. Then as the concentration falls the irregular and unreproduc-
ible effect of viscous fingering shows up, much more dramatically
here than in the usual chromatogram. The same two concentrations are
shown in frontal analysis in Figure 4. Here the curves are regular,
the frontal peak is delayed more than the small-sample peak, but the
return to baseline is earlier, and the curve shape is subtly altered.
The viscous fingering effect is not disturbing the elution curve as
in the small-sample case, but in Figure 5 it is shown in the flushing
curve, where the concentration falls from 10 mg/ml down to pure solvent.

Taking up the separation of a mixture, in Figure 6 lower concentrations, 0.25 mg/ml, are shown separately and together for polystyrenes of 2.15×10^6 and 4.11×10^5 M_w, weight-average molcular weight.
The frontal chromatograms are regular, the load effect is small and
the mixture is clearly resolvable. But in Figure 7, with 1.0 mg/ml
of each the compression delay is clearly seen. Figure 8 shows the
frontal chromatograms for a much broader sample, a mixture of four
standard polystyrenes run in three concentrations, and we will take
that up in greater detail. The first peak shows no load effect,
since it comes at the exclusion limit of the column. The first and
second peaks are the same polystyrenes used in Figures 6 and 7, while
the molecular weights of the third and fourth peaks are 160,000 and
51,000, respectively. The concentrations start at 1 mg/ml each
component in the top curve and go down by factors of four to 0.25
and 0.0625 mg/ml, respectively. In the middle chromatogram the load
effect is clearly discernible, even at a total sample concentration

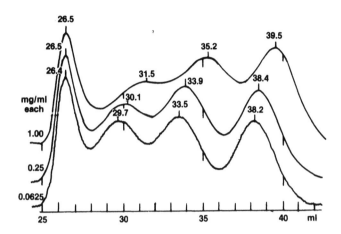

FIGURE 8. Frontal analysis of broad mixture at three concentrations.
Materials, readings and calculations in Table 1.

of only 1 mg/ml, or 0.1 percent. Here the delay effect, as the
flowing concentration, is accumulative so that the smaller polymers
are compressed the most in frontal analysis. But with viscous fingering absent we can attempt to derive a correlation for the compression effect.

Since this load effect is measured in the pores of the column,
it was easy to conceive of it as a compression, and an empirical
approach was tried first. Using the sum of the concentration x
intrinsic viscosity term as the pressure, an empirical expression
was formulated to derive the compressibility term from the intrinsic
viscosity [11]. For demonstration this worked fairly well, but a
more general approach was still needed. Our approach was then up-
graded in two ways.

First, we recognized that for short columns the gaussian curve
shape is a poor fit, and for mixtures of the narrow standard polystyrenes this became important, so we looked for more flexible forms.
Let us first write the gaussian equation in our terms. Where curve
height at volume V is H_v, at the peak it is Hp at V_p. With standard

σ we have

$$H_v = H_p e^{-((V - V_p)^2/2\sigma^2)} \qquad (1)$$

Since

$$H_p = \text{area}/\sigma\sqrt{2\pi} \qquad (2)$$

in the gaussian curve, for skewed forms we may use

$$\sigma = \text{area}/Hp\sqrt{2\pi}. \qquad (3)$$

An equation derived by Hess and Kratz [12] from packed bed mechanics was simplified to the form

$$H_v = H_p(V_p/V)^{1/2} e^{-((V - V_p)^2(V_p/V)^1/2\sigma^2)} \qquad (4)$$

This equation produced tailing sufficient for short columns with small molecules, but the pore diffusion problems of large molecules went of course beyond its premises. With V_p/σ low, it also showed the trait of giving H_v higher values than H_p just before V_p, so we decided to explore this form further. It was then generalized using exponents X, Y instead of the 1/2, 1 in Equation 4. By numerical integration we found that the correct area was obtained on a line with X, Y values of -1, 0; 0, 2/3; and 1/2, 1. A family of skewed curves was thus available, all with height going to zero at V = 0. The equation with X, Y = -1, 0 gave at large skewing a small hump later than V_p, while the values 0, 2/3 retained the peak at V_p while giving an excellent match in shape to our narrowest polymer elution curves. In our experimental use we found this equation very satis-factory, with the option of increasing the dissymmetry by artifi-cially advancing the sample injection point by a fictitious volume V_f. With this curve shape in our GPC correction program we were able to resolve the test mixtures quite well at low loads, even with the one-foot columns.

$$H_v = H_p e^{-((V - V_p)^2((V_p - V_f)/(V - V_f))^{2/3}/2\sigma^2)} \tag{5}$$

The second way in which we attempted to upgrade our correction of frontal chromatograms was with the Maron-Reznick equation for correcting apparent molecular size to zero-concentration size. In Reznick's thesis [10] the relation is given for polymers in good solvents

$$1/\varepsilon = sv + 1/\varepsilon_0 \tag{6}$$

This is the equation of a straight line in $1/\varepsilon$ and v, where $1/\varepsilon$ is the intercept and s is the slope. Here ε is a molecular volume factor in solution, and v is the volume concentration of the dry polymer in ml per ml of solution. He found a simple relation between this molecular volume factor and intrinsic viscosity, with polymer density d in g/ml.

$$\varepsilon_0 = 50 \ d[\eta] \tag{7}$$

The molecular volume factor at infinite dilution is thus ε_0, and the minimum value ε_∞ he found to be 4 to 5 regardless of polymer molecular weight, at 20 to 25 volume percent concentration. The slope s amounts to a compressibility index, where

$$s = (\varepsilon_0 - \varepsilon_\infty)/\varepsilon_0. \tag{8}$$

The hydrodynamic volume $[\eta]M$ is well supported as the effective measure of molecular size in GPC. Where load effects delay elution volume our calibration curve will give us an apparent $[\eta]M$ where, since the M is not compressible, the $[\eta]$ must be. Analogous to (7) I assume we may write

$$\varepsilon = 50 \ d[\eta]_{app}. \tag{9}$$

Since in the empirical correlation [10] I had used the specific viscosity $[\eta]_i c_i$ as the intermolecular pressure, as an assumption for study I took Equation 6 in our context as

$$1/\varepsilon = 1/\varepsilon_0 + \Sigma \ s_i v_i \tag{10}$$

Substituting (7) and (9) in Equation 10 we can solve for either the apparent or the corrected intrinsic viscosities

$$[\eta]_{app} = 1/(1/[\eta] + 50 \ d \ \Sigma \ s_i v_i) \tag{11}$$

$$[\eta]_{cor} = 1/(1/[\eta]_{app} - 50 \ d \ \Sigma \ s_i v_i) \tag{12}$$

Here Equation 11 is useful in a test situation, but Equation 12 is needed in an iterative program for zone broadening and load correction. Although the quantities s_i need $[\eta]$ the initial value can be $[\eta]_{app}$ and the program will converge as the corrected value replaces it.

The test situation shown in Figure 8 was run on a 3/8 inch × four foot column in three concentrations in tetrahydrofuran at room temperature. The flow rate was one ml per minute. The differentiating loop in the refractometer cell box was one meter of 1/16 inch stainless steel tubing having 0.67 ml internal volume. Since the first peak emerged at the interstitial volume, and no concentration effect was observable, this is evidence that we are dealing with a response of molecular size to concentration, in contrast to viscous fingering which is an effect of fluid flow outside the porous particles. Table 1 shows the data from Figure 8 plus another sample, the same first three polymers with an even lower fourth one. The peak volumes were read off and compared with apparent hydrodynamic volumes from column calibration data. The observed compression then is the ratio of the observed value to the calibration value of hydrodynamic volume, and this is compared with the calculated value obtained from Equation 11. However, since the compression effect was far too great the first time through, the calculation is shown with the compressing factor $\Sigma \ s_i v_i$ multiplied by 0.14, an unexplained "fudge factor" for your consideration.

A sample calculation follows: From Figure 8 the third peak with the 1 mg/ml sample concentration shows an elution volume read into Table 1 as 35.2 ml. This is a polystyrene of $M_w = 1.60 \times 10^5$, $[\eta]_{THF} = .641$ dl/g for a log $[\eta]M = 5.0112$ at 33.40 ml. With the curve essentially straight in this area we have at 35.2 ml

TABLE 1

FRONTAL ANALYSIS OF BROAD MIXTURES

Polystyrene #	P61970	4190037	41984	4190041	4190042
Mol Wt M_w	2.15×10^6	4.1×10^5	1.60×10^5	5.10×10^4	1.03×10^4
$[\eta]$ in THF, dl/b	4.93	1.24	0.641	0.274	0.102
log $[\eta]M$ dl/mole	7.0253	5.7076	5.0112	4.1452	3.0216
V_p no load, ml	26.4	29.55	33.40	38.08	42.80
0.0625 mg/ml each:					
V_p observed, ml	26.4	29.7	33.5	38.2	--
log $[\eta]M$ apparent (at V_0)		5.680	4.992	4.130	--
observed compression $[\eta]_{app}/[\eta]$		0.95	0.96	0.97	
calc. compr.		0.925	0.935	0.960	
error %		-2.65%	-2.60%	-1.03%	
0.25 mg/ml each:					
V_p observed, ml	26.5	30.1	33.9	38.4	--
apparent log $[\eta]M$ (at V_0)		5.607	4.919	4.088	
obs. compr.		0.793	0.808	0.876	
calc. compr.		0.754	0.781	0.856	
error %		-4.92	-3.33	-2.25	
1.00 mg/ml each:					
V_p observed, ml	26.5	31.5	35.2	39.5	--
apparent log $[\eta]M$ (at V_0)		5.352	4.679	3.882	
obs. compr.		0.441	0.465	0.546	
calc. compr.		0.434	0.471	0.599	
error %		-1.48	+1.29	+9.63	
0.0625 mg/ml each:					
V_p observed, ml	26.5	29.8	33.5	--	4.28
apparent log $[\eta]M$ (at V_0)		5.662	4.992		3.022
obs. compr.		0.900	0.957		1.00
calc. compr.		0.925	0.935		0.984
error %		+2.78	-2.30		-1.6
0.25 mg/ml each:					
V_p observed, ml	26.5	30.1	33.9	--	43.0
apparent log $[\eta]M$ (at V_0)		5.607	4.919		2.960
obs. compr.		0.793	0.808		0.867
calc. compr.		0.754	0.781		0.941
error %		-4.92	-3.34		+8.54
1.00 mg/ml each:					
V_p observed, ml	26.6	31.6	35.2	--	43.5
apparent log $[\eta]M$ (at V_0)		5.338	4.679		2.855
obs. compr.		0.427	0.465		0.681
calc. compr.		0.434	0.471		0.800
error %		+1.64	+1.29		+14.9

$\log [\eta]M_{app} = 4.679$. The difference is $-.3322$, which is the log of the compression ratio, and $[\eta]_{app}/[\eta] = 0.465$.

Using Equation 11, we note that for polystyrene $d = 1.05$ g/ml, and that Reznick's v_i is concentration in ml dry polymer substance per milliliter of solution. Since we are using c_i in mg polymer per ml of solution, we have

$$v_i = c_i/(1000 \text{ mg/g} \times 1.05 \text{ g/ml}) = c_i/1050 \qquad (14)$$

Substituting Equation 7 in (8) and using the given values 4 for ε_∞ and 1.05 for d, the expression reduces to

$$s_i = 1 - 0.07619/[\eta]_i, \text{ with } [\eta] \text{ in ml/g.} \qquad (15)$$

In Equation 11 the term $50d \Sigma s_i v_i$ then becomes $0.14 \times 50 \times 1.05/ 1050 \Sigma s_i c_i$, where c_i is the mg/ml of each polymer component in the solution at the given location. Since we needed a fudge factor, we make no claim that our use of the Maron-Reznick relation expresses all that is going on in frontal GPC, and we hope only to show a correlation that may be useful and provocative of further study. We now have

$$[\eta] \text{ apparent} = 1/(1/[\eta] + 0.0070 \Sigma c_i (1 - .07619/[\eta]_i) \qquad (16)$$

In this case, at $V_e = 35.2$ ml, for each component $c_i s_i =$
a. 1 mg/ml $(1 - .07619/493) = 0.99984$
b. 1 mg/ml $(1 - .07619/124) = 0.99939$
c. 0.5 " $(1 - 0.7619/64.1) = \underline{0.49940}$

$\quad [\eta]_{app} = 1/(1/64.1 + .0070(2.49863) = 30.22$

$\quad$ compression $= 30.22/64.1 = 0.471$, error $= 100(.471 - .465)/.465 = +1.3\%$.

The errors, of course, may be in either the calculated or the observed values.

V. CONCLUSIONS

We have shown that very short columns can produce interesting results in relatively short time, but as required by flow and diffusion mechanics, the single species elution curves will be more unsymmetrical than those at higher elution volumes. A curve equation in gaussian form is offered that is able to reproduce such shapes, and it approaches the symmetrical form as the ratio of elution volume to standard deviation becomes large. With frontal analysis the two load effects of viscous fingering and macromolecular compression were separated. Since both effects become large at about the same concentrations, it appears that frontal analysis is not a simple way to avoid the erratic disturbing effect of viscous fingering. The compression effect was shown to be regular and predictable, so that its correction may reasonably be incorporated in a correction for zone broadening.

ACKNOWLEDGMENTS

Thanks are due to Dow Chemical USA for permission to publish this work.

REFERENCES

1. Present address 422 Kent Avenue, Lake Placid Fla. 33852.

2. J. C. Moore, Separation Sci., 5, 723 (1970).

3. G. E. Hibbard, A. G. Ogston, and D. J. Winsor, J. Chrom., 48, 393 (1970).

4. W. H. Cheetham and D. J. Winsor, J. Chrom., 48, 400 (1970).

5. J. A. Ronalds and D. J. Winsor, Arch. Biochem. Biophys., 129, 94 (1969).

6. E. K. Bogacheva, A. V. Kiselev et al., Vysokomol. Soedin. Ser A, 11, 2180 (1969) (C.A. 72 #32380, transl.).

7. K. Hellsing, J. Chrom., 36, 170 (1968).

8. S. H. Maron, N. Nakajima, and I. M. Krieger, J. Polym. Sci.,
    37, 1 (1959).

9. R. B. Reznick, "The Flow Behavior of Dilute Polymer Solutions",
    Div. Macromol. Science, Case Western Reserve University, 1970.

10. R. B. Reznick, "The Flow Behavior of Dilute Polymer Solutions",
    Div. Macromol. Science, Case Western Reserve University, 1970,
    pp. 6-7, 59-72.

11. U. S. Patent 3,649,200 (March 14, 1972).

12. M. Hess and R. F. Kratz, J. Polymer Sci. A2, 4, 731 (1966).

THE USE OF GEL PERMEATION CHROMATOGRAPHY TO DETERMINE THE STRUCTURE AND POLYMERIZATION MECHANISM OF BRANCHED BLOCK COPOLYMERS

Michael R. Ambler

The Goodyear Tire and Rubber Co.
Akron, Ohio

ABSTRACT

The polymerization of styrene-diene branched block copolymer containing divinylbenzene can be followed by GPC. By combining responses of ultraviolet and refractive index detectors, the compositions of the various molecular weight species are determined. By combining intrinsic viscosity with the universal calibration of the GPC, extent of branching and molecular weight can be quantitated. By comparing these data for samples with different histories, e.g., a conversion study of one batch polymerization, the mechanism of formation of branched polymer structure can be determined. Using GPC, polymer structure can be monitored, as well as correlated to and optimized for final physical properties.

Examples of the terpolymerization of styrene (S), butadiene (B) and divinylbenzene (DVB), and styrene (S), isoprene (I) and divinylbenzene (DVB), to form randomly branched block copolymers of the general structure, [S(B-DVB)]x, and [S(I-DVB)]x, are presented. A few results of analyses of other types of branched and linear block copolymers are shown to illustrate the utility of the method.

I.   INTRODUCTION

The determination of random branching by GPC has developed into one of its more important applications. It has been observed, particularly for polymers susceptible to branching via side reactions when they contain multifunctional monomers like divinylbenzene or dienes

like butadiene and isoprene, that branching is as much an influence
upon polymer properties as is its molecular weight (MW or M) and
molecular weight distribution (MWD).

Various branching parameters can be calculated from a combination of intrinsic viscosity and GPC data [1-3]. Only a brief review
of the more important equations is made here. Refer to the original
articles for a more complete discussion. It is assumed in the work
presented here that the universal calibration is valid for branched
polymers. This appears not to be the case for severely branched
polymers [4], but for diene polymers with microgel and lower levels
of branching, universal calibration appears valid. The defining
equation,

$$\frac{[\eta]_b}{[\eta]_1} = h^3 \tag{1}$$

where $[\eta]_b$ and $[\eta]_1$ are the intrinsic viscosities of branched and
linear polymer molecules of the same structure and MW, and, h^3 is a
function of g, the so-called branching function, expresses the
change in hydrodynamic properties due to branching. Using Equation
1, several random branching parameters are calculated. Since g is
a function of m, the number of branch points, a branching density,
λ, is expressed as

$$\lambda = \frac{m}{M} \tag{2}$$

It is commonly assumed, and most literature work has shown this to
be accurate or at least reasonable, that λ is a constant over the
entire MWD for randomly branched polymer of broad MWD, i.e.,

$$\lambda \cong \overline{\lambda} \tag{3}$$

Assuming the polymer in question is broad in MWD and has a GPC curve
resembling that of Figure 1, the calculation generates a value of $\overline{\lambda}$
which in turn is used to calculate three random branching parameters:
(a) $\overline{M}_{bp}$, the MW between branch points for the highly branched molecules in the MWD; (b) M^*, the lowest MW species in the MWD that is
branched enough to reduce intrinsic viscosity; and (c) "% Branched,"

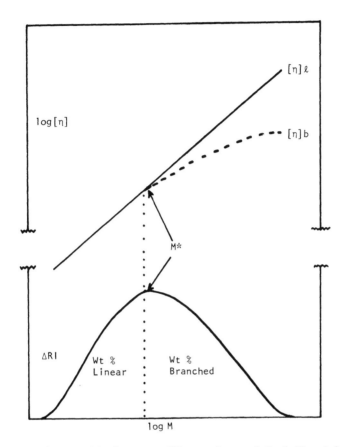

FIGURE 1.   Relationship between GPC results and Mark-Houwink plot for branched polymers.

the percentage of the polymer that is branched, as shown in Figures 1 and 2:

$$\overline{M}_{bp} = \frac{M}{(f-1)m} \tag{4}$$

$$M^* = \frac{f}{(f-1)(2\overline{\lambda})} \tag{5}$$

where f is the functionality of the branch point, equal to 4 in this study of the branching of tetra-functional DVB.

As can be seen in Figure 1, the [η]-M relationship for the branched polymer is a function of the Mark-Houwink coefficients K

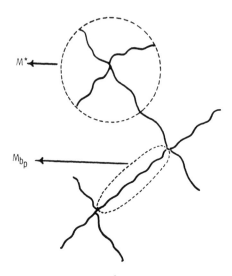

FIGURE 2.   Branching parameters M* and $\overline{M}_{bp}$.

and α for the linear species of equivalent structure with a correction for the reduction in [η] due to branching. In this article,
branching of S-B and S-I type block copolymers are being evaluated,
and the determination of K and α for these block copolymers in good
solvents must be described. Generally, it is found that, whether
plotting $\log^{[\eta]}$ or $\log R_G^2$, the square of the radius of gyration,
versus log M, one gets straight lines for two homopolymers, for
example, polystyrene and polybutadiene, while the same data for
copolymers of intermediate composition will fall in-between the
lines and shift more or less directly as the composition changes.
For example, in Figure 3 the Mark-Houwink relationships for polybutadiene of low vinyl-1,2 content, 25% styrene random S/B copolymer,
50% styrene S-B-S linear tri-block copolymer, and polystyrene, all
determined for THF at 25°C, illustrate this quite well. These lines
were determined experimentally either in our laboratory or taken
from the literature. Interpolation of both K and α as functions of
composition is possible, particularly in a good solvent where both
α's will be similar, approximately 0.7. A search of the available
literature values [5] has found that plots of log K versus weight

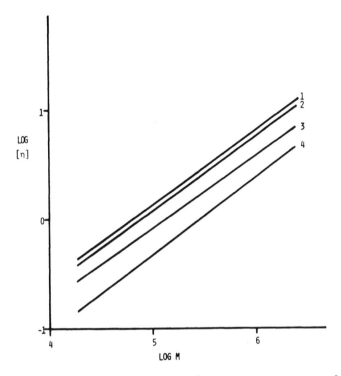

FIGURE 3. [η]-M relationships for S/B copolymers, THF, 25°C: 1,
polybutadiene, approx. 20% vinyl-1,2; 2, 25% bound styrene, random
S/B copolymer; 3, 50% bound styrene, S-B-S linear tri-block copolymer; 4, polystyrene.

percent (W) of one comonomer, and α versus W of one comonomer,
generates more or less straight or smoothly curved lines for random
copolymers. This empirical interpolation is also suggested and the
data appear feasible for this, when studying the change in K and α
for polybutadienes of different vinyl-1,2 contents [6]. Figures
4 and 5 illustrate that the S/B system in THF can be interpreted in
this way. Both random and block S/B copolymers describe the same
smooth progression in K and α with styrene content. If sufficient
data are available, the solid lines can be drawn. However, when
only data for the homopolymers are available, as it is here in this
work, linear relationships can be made, as illustrated by the dotted
lines. Note that in both figures the ordinate is highly expanded

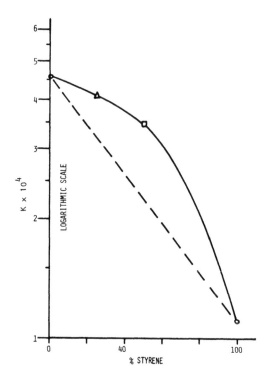

FIGURE 4. Plot of Mark-Houwink coefficient K versus percent bound styrene for S/B copolymers, THF, 25°C: o, polybutadiene, approx. 20% vinyl-1,2, and polystyrene; Δ, 25% bound styrene, random S/B copolymer; □, 50% bound styrene, S-B-S linear tri-block copolymer.

and any errors introduced by using the dotted line to establish K and α is small and will not seriously jeopardize the results, particularly since an underestimation of one parameter usually coincides with an overestimation of the other. As suggested indirectly by Chang [7] and Tung [8], and by Kuo's data [9], a similar interpolation of data in chloroform is considered sufficiently accurate for S-B and S-I block copolymers for use in this work. Having determined K and α for monodisperse polystyrene, polybutadiene and polyisoprene polymers, these parameters were plotted linearly as log K versus W and α versus W for the two polymer families. Relevant K and α values were then interpolated from these graphs for each sample of known comonomer ratio. This technique of estimating

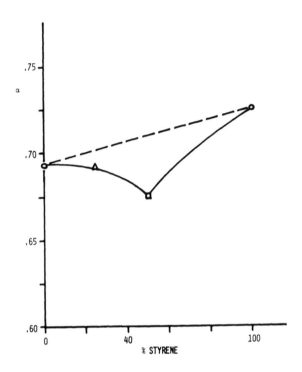

FIGURE 5. Plot of Mark-Houwink coefficient α versus percent bound
styrene for S/B copolymers, THF, 25°C: o, polybutadiene, approx.
20% vinyl-1,2, and polystyrene; Δ, 25% bound styrene, random S/B
copolymer; □, 50% bound styrene, S-B-S linear tri-block copolymer.

Mark-Houwink coefficients will not apply to all block copolymer
systems [10], but our data suggest that this assumption does not
seriously detract from what we are attempting to accomplish.

The use of the dual detectors ultraviolet (UV) and refractive
index (RI) allows the composition of each monodisperse increment of
elution volume to be followed if the two comonomers have different
relative responses [11]. Since styrene absorbs strongly at 262 nm,
while butadiene and isoprene absorb weakly, and since all three
monomers have similar RI responses, the UV/RI detector response ratio
is proportional to weight percent styrene W(S),

UVαC(S)   (6)

$$RI \alpha C(S + B) \tag{7}$$

$$\frac{UV}{RI} \alpha \frac{C(S)}{C(S + B)} \alpha W(S) \tag{8}$$

where the notation $C(X)$ represents concentrations. Thus, by ratio-
ing the recorder heights of the UV and RI responses at constant
elution volume, a value roughly proportional to $W(S)$ is calculated.
By comparing this ratio at different elution volumes, changes in
styrene content can be sensed without generating actual weight
percentages.

One always runs the risk, when evaluating polymer systems capable of branching, of creating gel. This can be conveniently isolated
quantitatively from the portion of the polymer to be analyzed [1].

Rubber-resin block copolymers like S-B or S-I copolymers are
thermoplastic in nature, and between the two Tg's of the different
block segments they are highly elastic and have very high strength.
It is critical in obtaining the highest strength possible that only
tri-block S-B-S or S-I-S species are present so that each molecule
is permanently anchored in two different styrene domains and high
elasticity is achieved. Homopolystyrene impurity usually affects
physical properties, and di-block S-B or S-I impurity does not contribute efficiently to elasticity since it is not physically
anchored at two points and is a network defect. Thus, even in a
simple linear block copolymer, it is important to control and maxi-
mize the tri-block concentration. In the more complex branched
copolymers, the more styrene segments a molecule contains, the
greater the number of separate styrene domains the molecule will
anchor into, and strength can be further increased. Here, even a
linear tri-block species could be regarded as an impurity of perhaps lower strength than the product. Thus, it is also important
in branched copolymers to control and optimize the concentration
and character of the branched tri-block species.

The preparation of randomly branched block copolymers of S and B
or S and I, each containing small amounts of DVB to act as tetra-functional branching agent, can be prepared by an anionic polymerization

process [13]. It is the purpose of this article to illustrate how
GPC data and branching calculations can be used to help determine
this polymerization mechanism. Other data are needed to supply a
complete picture of the mechanism and structure, including monomer
depletion curves, spectroscopic analyses and physical property
measurements, but this article will not deal with these data.

II. EXPERIMENTAL

A Waters Model GPC-100 was operated on chloroform at 30°C. Practical grade solvent had been distilled from glass and spiked with 1/2%
each of mechanol and 1-hexene inhibitors. The standard RI detector
was followed in series with a Varian Model 635 UV/Visible Spectrophotometer equipped with a flow cell. Absorption at 262 nm was monitored.
The flow rate was 1 ml/min. Various combinations of 4 foot long
Styragel columns with no gaps in porosity were used to ensure linear
calibration over the elution volume of interest. At all times at
least seven columns were used. Calibration was made with polystyrene
standards available from The Goodyear Tire and Rubber Company (Chemical Development Services, Akron, Ohio 44316). Commercial block
copolymers were obtained from Phillips Petroleum Company (Bartles-
ville, Oklahoma) and Shell Chemical Company (Houston, Texas).
Intrinsic viscosities were determined in chloroform at 30°C using a
low shear Ubbelohde capillary viscometer ($\dot{\gamma}_w < 100$ sec^{-1}). Linear
plots of ηsp/C versus C (gm/dl) were used to determine [η]. All
solutions were filtered through 0.45 micron porosity Gelman membranes
of suitable chemical compatibility. All calculations were done by
computer using a proprietary computer program. The Perkin-Elmer
Model 283 Infrared Spectrophotometer was used to determine W(S) in
the various samples. Spectra of CS_2 solutions were analyzed using
calibrations based upon polystyrene and verified as accurate by
nuclear magnetic resonance analyses.

The polymerization will be described only in general terms [13].
In one vessel, purified styrene is added to purified inert solvent,

and then the initiator, an alkyllithium, is added. Polymerization
of polystyryllithium is allowed to occur to complete conversion of
monomer to polymer. In a second vessel, liquified butadiene or
purified isoprene is dissolved in purified solvent, purified DVB is
then added, and a small amount of initiator is added to the second
vessel and polymerization is allowed to proceed. After complete
conversion of monomer to polymer, the solution is aged further and
then a methanol solution of phenolic antioxidant is added to stop
the reaction. The polymer is then isolated and dried.

III.   RESULTS AND DISCUSSION

A.   Qualitative Analyses

The manner in which these polymers are made suggests that the resultant polymers will be branched through the DVB. But how? And what
branched structure is formed?

The first experiment involved the GPC analysis of three samples
(B-473, B-476, B-477) of S(B-DVB) terpolymers. Their GPC curves are
sketched in Figure 6. These are representative of the possible
types of polymer that can be made. The polymers exhibit at least
three peaks. The peaks at around counts 125-130 were verified as
the homopolystyrene end-block by GPC analysis of this part of the
multi-step polymerization. The extent and MW of "homopolystyrene"
can vary depending upon polymerization conditions. The higher MW
peaks, counts 105-118, correspond to S(B-DVB) type species of varying MW's, degrees of branching, composition (di-block versus tri-
block) and microstructure (weight percent styrene in each block
species). As will be evident later, the same type of polymer profile is found with the S(I-DVB) system. The three samples in
Figure 6 represent final material obtained at 100% monomer conversion. It is obvious that a rather complex structure is obtained.
To learn how this complex structure is formed, a brief study of the
kinetics of the polymerization was made.

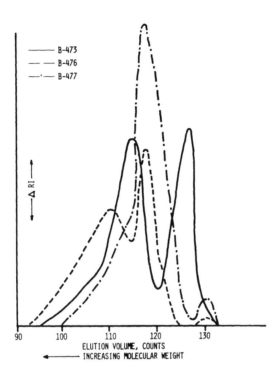

FIGURE 6.   GPC curves of $[S(B\text{--}DVB)]_x$ branched block copolymers.

During the polymerization of B-476, samples were removed from
the reactor and analyzed by GPC. The GPC curves are shown in Figure 7. Virtually 100% monomer conversion had occurred after 18
minutes polymerization time, but the apparent MW continued to grow.
This was felt to be the branching reaction of DVB units between
chains. Those of the higher sampling times of 34 to 90 minutes,
and at 120 minutes, the final product, were virtually identical.

Two different S(I-DVB) terpolymerizations (I-106 and I-109)
were also sampled and evaluated. This time the GPC curves (not
shown) consisted of only two peaks, one the homopolystyrene, the
other presumably the copolymer product. For both samples, the peak
elution volume of the copolymer product was found to decrease with

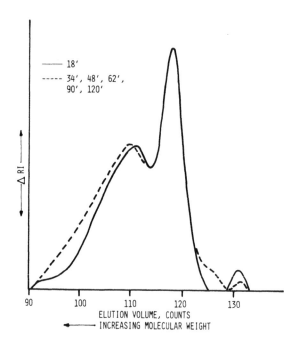

FIGURE 7.   Sample B-476:   GPC curves of one [S(B-DVB)]$_x$ branched
block copolymer sampled at different polymerization times.

TABLE 1

Polymerization Kinetics of S(I-DVB)

Sample	Reaction Time (min.)	Peak Elution Volume (counts)
I-106	45	88.0
	60	87.0
	90	85.5
	120	84.2
	165 (final product)	83.9
I-109	17	88.0
	21	86.0
	41	84.0
	45 (final product)	83.9

reaction time, indicating an increase in MW and/or branching as the
polymerization proceeded. This is summarized in Table 1. The monomer had been largely (around 90%) converted to polymer by the sampling times listed, consequently the GPC changes represent predominantly changes in branching with some additional minor MW increases due
to monomer conversion.

These preliminary experiments serve to illustrate the type of
polymer that is produced using this polymerization mechanism.
Obviously, the polymer structure is highly complex. Several questions need to be answered to adequately characterize the polymer.
For each of the various GPC peaks, we need to know (a) its comonomer
composition; (b) its weight percentage of the whole polymer; (c) its
MW; (d) its block nature, that is, is it di-block or tri-block; (e)
its type of branching, that is, linear, comb-, graft-, star- or
random-branching; and (f) its extent of branching, as characterized
by $\overline{M}_{bp}$ and M^*. Experiments were designed to obtain this information.

B. Compositional Analysis

Another sample, B-24, was chosen for detailed analysis because of
its similarity to those samples shown in Figures 6 and 7 with tri-
modal MWD's. The sample contained no gel and the GPC chromatogram
(Fig. 8) indicated a small homopolystyrene peak. Two traces are
shown for the dual detector system in chloroform, RI and UV at 262 nm.
Three peaks were found. The peak at 100 counts was found to be polystyrene. Using the $M[\eta]$ universal calibration, the MW of the
homopolystyrene was round 15,000, which was the same as that previously determined for the first polystyrene block of the S(B-DVB)
polymerization. A calibration curve of peak height (RI detector)
versus concentration was generated for polystyrene of the same MW
and used to determine the weight percentage of the homopolystyrene
in the sample. The sample was found to have 2.5% homopolystyrene.
The precision of this measurement was about ±3% of the value. Infrared analysis of the whole sample indicated the bound styrene to be

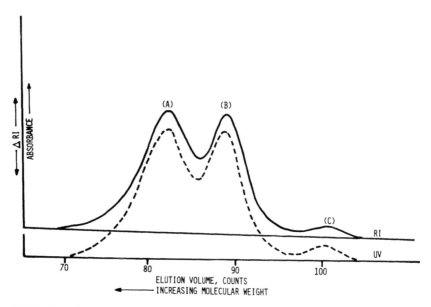

FIGURE 8. Sample B-24: GPC curves from dual detectors of one $[S(B-DVB)]_x$ branched block copolymer.

34.3%. After correcting for the homopolystyrene content, the average bound styrene for the two S(B-DVB) species, that is, the two peaks A and B in Figure 8, was calculated to be 32.6%. In order to determine the composition of each of the two peaks, the responses from the dual detectors were correlated. Styrene and butadiene (and also DVB) have different, but fairly similar, refractive index detector responses, that is, refractive index increment or dn/dc. On the other hand, styrene and DVB exhibit a much stronger UV absorption at 262 nm than does butadiene (although it has been observed that anionically polymerized butadiene has an unexpectedly high UV absorption). Therefore, the ratio of UV response to RI response was used as a measure of the styrene and DVB contents. In this case, as the ratio of the UV/RI responses increases, the sum of the styrene and DVB contents also increases. As seen in Table 2, the UV/RI ratio for the peaks A and B are virtually identical and are less than that of the polystyrene peak C. It was deduced from this that both peaks

TABLE 2

GPC Data of Sample B-24

	Whole Sample	Peak A	B	C
Peak Elution Volume, counts	--	82.0	88.7	100.5
UV/RI (cm/cm)	--	1.067	1.064	1.419
Percent Styrene, %	34.3	32.6	32.6	100.0
M[η] (a)	--	942,000	110,000	2,000
Percent of Whole Sample, %	100.0	<—— 97.5 ——>		2.5
Molecular Shapes, MW[a]				
Linear	--	362,000	107,000	15,000
Tri-functional star-branched	--	> 362,000	110,000	--
Randomly branched, f = 4, m = 2	--	> 362,000	118,000	--
Model Species, MW[a]				
Linear Di-block, S(B-DVB)	--	--	46,900	--
Linear Tri-block, S(B-DVB)$_2$	--	--	93,800	--
Branched Tri-block, S(B-DVB)$_3$	--	--	141,000	--

[a]Polystyrene: $\alpha = 0.743$, $K = 1.01 \times 10^{-4}$.
Linear 33% S-B: $\alpha = 0.755$, $K = 1.65 \times 10^{-4}$.

were S(B-DVB) species and each had the same composition, 32.6% styrene. The percentage of DVB in these samples is negligible compared to styrene. This suggested that the branching reaction could involve successive coupling of di-block S(B-DVB) species to an [S(B-DVB)]$_x$ structure.

There is no direct way to calculate the MW of peak B since all that is known about peak B is its value of M[η] at its peak elution volume, determined from the universal calibration. One way to deduce its structure is to take an indirect approach, that is, to assume various linear and branched models (consistent with the possible ways they could be formed in the polymerization) and calculate the MW that each model would require to elute at the position of peak B. If the species of peak B were a block copolymer of linear structure, its MW would be calculated from the Mark-Houwink relationship as 107,000. This is the lowest MW value it could be and still have 32.6% styrene, since any branched species eluting at peak B would be of higher MW. For example, a tri-functional star-shaped molecule of 32.6% styrene

would have a MW of 110,000, and a randomly branched structure with
two tetra-functional branch points would have a calculated MW of
118,000. Consider now the theoretical MW of the linear di-block
S(B-DVB) sub-unit that is formed. Its styrene segment has a MW of
15,000 and its composition is 32.6% styrene. Therefore, the calculated MW for the linear di-block S(B-DVB) repeat unit

(I)

is 46,900. Of course, Structure I is possible only if the DVB units,
which are presumed distributed throughout the "B + DVB" rubbery segment, had not branched inter-molecularly or intra-molecularly through
its pendant vinyl group. If branching had occurred, the simplest
structure would be if one DVB unit had branched into two chains,
coupling them, each having the same MW as (I). Coupling could be
either at the end of the rubber segments (Structure II) or in the
middle of one or both rubber segments (Structure III). In both
cases, the overall MW is 93,800. This is in reasonable agreement
with the predicted MW of peak B if it were linear copolymer. (The
nature of the branch point, whether it be two DVB pendant vinyl
groups or one DVB and one double bond of Bd, or the DVB pendant vinyl
group actually polymerized into the second chain, is unknown and not
germane to this study since it is confined to evaluation of hydrodynamic properties.) Thus, peak B is undoubtedly a tri-block species
composed of two chains, and is probably more or less linear, depending on whether the active lithium end couples with a DVB unit near
the end of the second chain

(II)

or near the center of the rubbery block

(III)

The data rule out peak B as being a larger, 3-chain coupled, molecule since its MW, 141,000, is much too large for any of the possible branched structures listed previously (110,000 or 118,000) that it would have to have. (These arguments are based strictly on the peak elution volume of peak B. There probably is some di-block species of 46,900 MW present, since it should elute at around 92 counts. Also, peak B appears to be broader in MWD than expected for a monodisperse species like peak C. Undoubtedly, blends of species like (II) and (III) are present, besides species with more than two chains coupled.) That is, a tri-functional branched species of 141,000 MW would elute at an elution volume much lower than that of peak B.

The problem of assigning MW and structure to peak A is virtually impossible solely from these data. It undoubtedly has a higher MW based on its elution volume. However, universal calibration may no longer be valid, that is, the value of $M[\eta]$ of 942,000 may be too low [4]. If it was a linear molecule of 32.6% styrene, its MW would be 362,000, suggesting at least nine chains coupled, but since a branched structure is much more probable with nine or more chains, the MW is estimated to be greater than 362,000, perhaps by a factor of two or more. Beyond this assignment, no MW's can be determined. As to the structure of peak A, the first inclination is to rule out a linear structure since it is almost impossible to assign a structure with nine or more chains where they could combine to form a more or less linear structure. The manner of polymerization apparently will not allow this structure. After growing S blocks, the polymerization proceeds via a copolymerization of B and DVB. The terminal lithium species can polymerize through the pendant vinyl groups of DVB units in other chains, producing branching. If DVB is distributed randomly along the B chain, a mostly randomly branched structure would form. On the other hand, if B preferentially homopolymerizes and forces DVB to homopolymerize at the end as a separate "third" block segment, DVB-rich chain ends are formed at high conversions which would favor star structures being formed. Depending on how

the two monomers relatively deplete during the polymerization, it
is conceivable that the final product could favor either a randomly
branched structure

(IV)

or a star-shaped structure whose legs are more lightly randomly
branched or even predominantly linear.

(V)

In all probability, peak A is a very complicated blend of these
different branched structures and also different MW's. This is
indicated by the broadness of peak A and the presence of the high
MW tail.

It appears that it is not likely that B-24 contains a large
amount of di-block structure. Di-block structure is known to be
highly deleterious to the tensile strength and physical properties,
presumably because the rubbery chains cannot be anchored in a second
glassy domain and cannot contribute to the strength of the system.
In the same manner, in $[S(B-DVB)]_x$ there are probably dangling un-
anchored rubbery sections which may act as flaws to the strength of
the system (Structures III, IV and V).

These data provide some clues as to the polymerization mechanism.
Due to the nature of anionic polymerizations to make narrow MWD polymers, the resultant randomly branched polymer can be viewed as a
series of coupled di-block S(B-DVB) species with species having MW's
that are multiples of the MW of the linear di-block precursor sub-
unit. Based on total bound styrene and the MW of the polystyrene
segment, the MW of all possible species can be calculated. The viewing of the final polymer's structure in this way does not necessarily

mean it is made in this way, although depending upon polymerization
and aging of the living polymer, it may or may not be.

Even without knowing branching structure to any finer detail
than this, certain measurements and calculations can be made to further quantify the extent of branching and better define the overall
structure of $[S(B-DVB)]_x$ and $[S(I-DVB)]_x$. This was the object of
further studies.

C. Quantitative Branching Analysis

To further evaluate branched $[S(B-DVB)]_x$, twenty-eight samples were
prepared and analyzed. The GPC curve consisted of three peaks, a
lower MW peak, the homopolystyrene impurity, and two peaks comprising the S-B copolymer species. These two S-B copolymer peaks re-
sembled in most cases an envelope peak, this envelope being bi-modal.
The bi-modality of the S-B envelope peak varied from two clearly
separated peaks to one peak with a weakly defined shoulder on either
the high or low MW side. The amount of homopolystyrene impurity was
calculated from the area of its peak. After determining the total
bound styrene content via infrared measurements, the styrene content
of the bi-modal S-B species envelope was then calculated from

$$(\% \ \text{Styrene})_E = \frac{(\% \ \text{Styrene})_{total} - W_s}{100 - W_s} \qquad (9)$$

where W_s is the weight percent of homopolystyrene in the sample.
Again it was found that all S-B species under the envelope were of
the same styrene content, after visually comparing the UV and RI
traces. Next, the MW, $\overline{M}_s$, and intrinsic viscosity, $[\eta]_s$, of the
homopolystyrene peak were calculated from its universal calibration
value, J_s, based on its elution volume on the GPC and appropriate
Mark-Houwink coefficients

$$[\eta]_s = K_s^{1/(1+\alpha_s)} \cdot J_s^{\alpha_s/(1+\alpha_s)} \qquad (10)$$

and

$$\overline{M}_s = (J_s/K_s)^{1/(1+\alpha_s)} \tag{11}$$

where

$$J_s = (M[\eta])_s. \tag{12}$$

Then, knowing $\overline{M}_s$ and $(\% \text{ Styrene})_E$, the MW of the S-B repeat unit (I), $\overline{M}_u$, was calculated from

$$\overline{M}_u = \frac{\overline{M}_s \cdot 100}{(\% \text{ Styrene})_E} \tag{13}$$

Note that $\overline{M}_u$ represents the MW of the S(B-DVB) copolymer that would be formed if only the first double bond of DVB reacted into the chain and the pendant vinyl group did not polymerize. In this case, DVB would be like styrene in its functionality, and the di-block species would be linear. All other branched species will have MW's that will be multiples of $\overline{M}_u$. Next, the intrinsic viscosity of the sample, $[\eta]_t$, was measured. From $[\eta]_t$, $[\eta]_s$ and W_s, the intrinsic viscosity of the S(B-DVB) envelope, $[\eta]_E$, was calculated from

$$[\eta]_E = \frac{100 \cdot [\eta]_t - W_s \cdot [\eta]_s}{1 - W_s} \tag{14}$$

In all cases, the resolution of the S(B-DVB) envelope peak of the GPC curve was sufficient to describe it as a sum of two separated Gaussian-shaped peaks of assignable peak elution volumes. A computer program was written, where from the shape characteristics of the envelope and the two peak elution volumes, the envelope was mathematically deconvoluted into two Gaussian-shaped peaks eluting at the assigned peak elution volumes. The program's output listed all the possible combinations (sometimes over a hundred), together with the computed areas of each peak and the calculated total area. Final judgement as to the best fit was based heavily on a comparison

of calculated and measured total area, followed by observation of
the generated peaks and a comparison of the calculated and measured
envelope curve shapes. This technique provided the weight percentages of the high, W_H, and low, W_L, MW peaks of the envelope, defined
relative to the envelope as

$$W_H + W_L = 100. \tag{15}$$

From the original GPC, overall bound styrene and intrinsic
viscosity data, the percentage and MW of homopolystyrene, the bound
styrene content of each S–B species, the theoretical MW's of linear
S(B–DVB) di-block precursor, linear [S(B–DVB)(DVB–B)S] tri-block
precursor, and higher order multiples of $[S(B-DVB)]_x$ tri-block
branched species, and the GPC curve and weight percentages of the
usually two types of S(B–DVB) species comprising the block copolymer,
all had been generated. All these data are summarized in Tables 3
and 4. What remains is to determine what structure and MW the two
peaks have that comprise the envelope peak. These subsequent MW and
branching calculations must be made assuming specific branching
models constrained only by these data and what knowledge we have of
the manner of polymerization. It is this information which allows
a determination of polymerization mechanism and provides parameters
for correlating molecular and physical properties.

First the low MW envelope peak was considered. Its simplest
structure would be linear S(B–DVB) di-block (MW equal to $\overline{M}_u$); its
next possible structure would have to be linear [S(B–DVB)(DVB–B)S]
tri-block (MW equal to $2\overline{M}_u$). In either case, from its Mark–Houwink
coefficients and GPC curve, its intrinsic viscosity, $[\eta]_L$, was
calculated, and the MWD of the curve defined from the calculated
$(\overline{M}_w)_L$ and $(\overline{M}_n)_L$ values. In this case, $[\eta]_L$ is summed over the MWD
of that peak and not calculated directly from $(\overline{M}_p)_L$. Results are
listed in Table 5. For comparison purposes, the calculated peak MW
of the curve was calculated as

$$(\overline{M}_p)_L = [(\overline{M}_w)_L \cdot (\overline{M}_n)_L]^{1/2} \tag{16}$$

TABLE 3

Analysis of S(B-DVB) Samples

Sample	W_s	(% Styrene)$_E$	$[\eta]_t$	$[\eta]_s$
7	1.5	34.0	1.300	0.255
10	1.8	35.9	1.168	0.131
13	1.3	33.9	1.20	0.152
20	4.8	31.1	1.260	0.148
23	2.2	32.8	1.100	0.114
24	2.5	32.7	1.354	0.146
28	1.5	26.4	1.310	0.115
29	2.3	34.8	1.510	0.157
30	1.7	31.4	1.680	0.161
31	2.5	30.9	1.550	0.181
32	3.2	30.5	1.950	0.198
33	4.0	33.3	1.990	0.236
34	3.4	31.6	1.780	0.219
36	2.6	31.7	1.380	0.174
41	3.3	30.8	1.240	0.194
42	1.0	27.1	1.179	0.122
43	2.5	31.3	1.452	0.172
47	4.2	30.8	1.870	0.258
56	3.1	31.5	1.560	0.180
57	4.0	30.9	1.390	0.199
60	3.2	32.4	1.400	0.175
64	3.2	33.0	1.390	0.226
65	5.5	29.8	1.690	0.172
66	1.7	29.7	1.490	0.156
67	2.2	30.0	1.580	0.172
68	2.3	33.0	1.490	0.199
69	2.5	33.3	1.590	0.204
71	2.2	32.0	1.560	0.233

TABLE 4

Analysis of S(B-DVB) Samples

Sample	$\overline{M}_s$	$\overline{M}_u$	$[\eta]_E$	W_H	W_L
7	37,900	112,000	1.320	39	61
10	15,500	43,300	1.187	86	14
13	18,900	55,700	1.214	75	25
20	18,200	58,500	1.316	41	59
23	12,800	39,100	1.122	62	38
24	18,000	54,900	1.385	59	41
28	13,000	49,200	1.328	62	38
29	19,800	56,900	1.542	63	37
30	20,400	64,800	1.706	70	30
31	23,900	77,400	1.585	61	39
32	26,900	88,400	2.008	34	66
33	34,100	102,000	2.063	7	93
34	30,900	97,800	1.835	8	92
36	22,700	71,500	1.412	14	86
41	26,200	85,200	1.276	40	60
42	14,000	51,800	1.190	54	46
43	22,400	71,600	1.485	54	46
47	38,600	125,000	1.941	67	33
56	23,600	75,100	1.604	33	67
57	27,100	87,800	1.440	72	28
60	22,800	70,400	1.440	16	84
64	32,200	97,900	1.428	20	80
65	22,300	74,800	1.778	68	32
66	19,600	65,900	1.513	45	55
67	22,300	74,500	1.612	39	61
68	27,200	82,400	1.520	39	61
69	28,100	84,300	1.626	27	73
71	33,600	105,000	1.590	40	60

TABLE 5

Analysis of Low Molecular Weight S-B GPC Peak

Sample	$[\eta]_L$	$(\overline{M}_w)_L$	$(\overline{M}_n)_L$	$(\overline{M}_w)_L / (\overline{M}_n)_L$	$(\overline{M}_p)_L / \overline{M}_u$
7	0.991	107,000	73,500	1.45	0.8
10	0.471	39,700	33,400	1.19	0.8
13	0.594	53,500	42,000	1.28	0.8
20	0.993	103,000	73,800	1.40	1.5
23	0.762	73,500	55,300	1.33	1.6
24	0.975	102,000	75,500	1.35	1.6
28	0.718	62,000	49,000	1.26	1.1
29	0.971	104,000	75,700	1.37	1.6
30	0.967	98,500	76,800	1.28	1.3
31	0.965	98,200	77,000	1.27	1.1
32	1.648	203,000	134,000	1.51	1.9
33	1.947	266,000	162,000	1.64	2.0
34	1.587	197,000	88,400	2.23	1.3
36	1.168	128,000	90,200	1.42	1.5
41	0.840	82,600	60,200	1.37	0.8
42	0.721	63,600	48,900	1.30	1.1
43	0.902	90,300	67,800	1.33	1.1
47	1.064	112,000	83,400	1.35	0.8
56	1.172	129,000	89,600	1.44	1.4
57	0.672	60,000	52,000	1.16	0.6
60	1.157	128,000	90,400	1.42	1.5
64	1.180	133,000	88,200	1.50	1.1
65	0.924	90,700	65,800	1.38	1.0
66	1.108	116,000	79,200	1.47	1.3
67	1.193	128,000	89,000	1.44	1.4
68	1.057	113,000	83,900	1.35	1.2
69	1.274	148,000	99,600	1.49	1.4
71	0.995	104,000	73,300	1.42	0.8

The values of $(\overline{M}_w)_L/(\overline{M}_n)_L$ ranged from 1.16–2.23, suggesting a relatively narrow MWD. Knowing $(\overline{M}_p)_L$ and $\overline{M}_u$, an estimate of the structure of the species is possible. Considering its definition, $\overline{M}_u$ would represent the lowest MW copolymer species possible. Higher MW species will only come about by branching through the pendant vinyl group of DVB. With this in mind, a comparison of the average MW of the low MW peak, $(\overline{M}_p)_L$, to that of $\overline{M}_u$, provides an idea of the structure of the peak. For example, if $(\overline{M}_p)_L/\overline{M}_u$ is equal to one, the species is linear di-block; if $(\overline{M}_p)_L/\overline{M}_u$ is equal to two, the structure is two coupled di-block units, a tri-block species of probably linear structure. If $(\overline{M}_p)_L/\overline{M}_u$ is greater than two, more di-block units are coupled to form branched tri-block species. The calculated value of $(\overline{M}_p)_L$ will be accurate only if the species is linear copolymer. If it is branched copolymer, its real MW will be even higher than that value. This would of necessity not allow further calculations since the structure and MW of the low MW peak of the envelope must be known accurately. However, this was not found to be a problem. When the values of the ratio $(\overline{M}_p)_L/\overline{M}_u$ were averaged for all 28 samples, a relatively constant value, 1.2 ± 0.4, was obtained. Considering the average value of 1.2, and the fact that the distribution of the peak is ≤ 2.2, this strongly suggests that the low MW peak is a mixed linear species of di-block and tri-block nature. The results indicate that the samples really are not much different with respect to the structure of the low MW copolymer species. That is, for each sample, we can regard the low MW envelope peak as being linear in structure, and depending on whether $(\overline{M}_p)_L$ approached $\overline{M}_u$ or $2\overline{M}_u$, di-block or tri-block in nature. This does not mean that the samples are similar in their content of low MW linear copolymer species, but only the absence of a predominance of branched species is indicated. In Figure 9 is plotted the values of $\overline{M}_u$ versus $(\overline{M}_p)_L/\overline{M}_u$ for all 28 samples. There appears to be no dependence in the di-block/tri-block nature of the low MW peak, that is, $(\overline{M}_p)_L/\overline{M}_u$, on the repeat unit MW, $\overline{M}_u$. Note that none of the samples appear to be branched, ratio > 2, and all ratios fall quite well within the range 0.8–2.0.

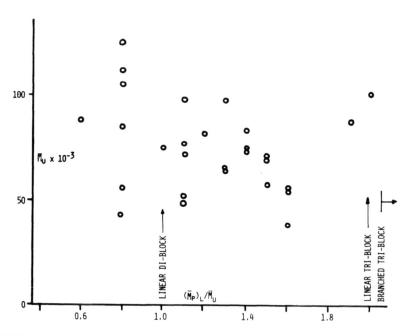

FIGURE 9. [S(B-DVB)]$_x$: Molecular weight relationship between
predicted di-block repeat unit and observed low molecular weight
copolymer species.

Calculations were now performed on the high MW envelope peak.
Results are shown in Table 6. Its intrinsic viscosity, $[\eta]_H$, was
calculated from

$$[\eta]_H = \frac{[\eta]_t - (W_s/100)[\eta]_s - (W_s/100)[\eta]_L + (W_s/100)(W_L/100)[\eta]_L}{(W_H/100) - (W_s/100)(W_H/100)} \quad (17)$$

Since it was felt that the high MW peak was branched tri-block species, the GPC and intrinsic viscosity data of this peak were combined with the appropriate Mark-Houwink coefficients and random
branching analyses were performed as described by Equations 2 - 5
and depicted in Figures 1 and 2. In Figure 10, it is seen that for
21 of the 28 samples, almost total branching extents were found,
while for 7 samples, very low branching was indicated. This indicates that, although all these samples had observable high MW peaks,
not all of these peaks were branched species. The significance of

TABLE 6

Analysis of High MW S-B GPC Peak

Sample	$[\eta]_H$	$(\bar{M}_{wb})_H$	$(\bar{M}_{nb})_H$	$(\bar{M}_{wb})_H/(\bar{M}_{nb})_H$	(% Branched)$_H$	$(\bar{M}_{bp})_H$	$(\bar{M}_p)_H/(\bar{M}_p)_L$
7	1.823	1,905,000	421,000	4.52	100	25,900	10.1
10	1.304	24,500,000	215,000	114.2	98	14,500	63.0
13	1.825	3,020,000	262,000	11.5	95	31,900	18.8
20	1.781	1,060,000	454,000	2.35	100	23,200	8.0
23	1.343	1,690,000	440,000	3.83	100	13,400	13.5
24	1.670	10,900,000	447,000	24.3	100	18,600	25.2
28	1.702	827,000	260,000	3.18	98	26,400	8.4
29	1.877	6,200,000	509,000	12.17	100	22,900	20.0
30	2.023	2,050,000	361,000	5.68	99	32,000	9.9
31	1.981	495,000	211,000	2.35	75	78,100	3.7
32	2.706	1,540,000	576,000	2.67	100	48,800	5.7
33	3.606	2,370,000	1,590,000	1.49	100	55,000	9.4
34	4.686	1,200,000	794,000	1.51	85	226,000	7.4
36	2.915	410,000	346,000	1.19	0	∞	3.5
41	1.929	232,000	176,000	1.32	0	∞	2.9
42	1.589	251,000	156,000	1.61	68	62,700	3.5
43	1.982	286,000	192,000	1.49	7	254,000	3.0
47	2.372	1,540,000	333,000	4.61	95	52,400	7.4
56	2.481	556,000	377,000	1.47	96	79,400	4.2
57	1.738	205,000	132,000	1.55	0	∞	2.9
60	2.930	416,000	347,000	1.20	0	∞	3.5
64	2.423	507,000	390,000	1.30	99	81,200	4.1
65	2.144	963,000	254,000	3.79	91	52,000	6.4
66	2.008	952,000	414,000	2.30	100	30,700	6.5
67	2.266	664,000	394,000	1.69	99	46,600	4.8
68	2.245	346,000	247,000	1.40	9	258,000	3.0
69	2.576	523,000	370,000	1.42	85	112,000	3.6
71	2.482	443,000	265,000	1.67	15	226,000	3.9

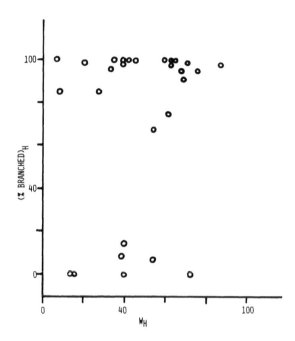

FIGURE 10. [S(B-DVB)]$_x$: Relationship between percentage of high molecular weight copolymer species that is branched and percentage of high molecular weight copolymer species in the sample.

this observation regarding the polymerization mechanism is unknown, but is not apparently related to the fact that in these samples, the lower MW envelope peak is the linear di-block and the higher MW peak is the linear tri-block. Figure 11 shows no strong relationship between $(\overline{M}_p)_L/\overline{M}_u$ and the percentage of the high MW peak that is branched. Here, the triangular data points are the seven lower data points of Figure 10. However, there is a suggestion that for those samples whose high MW peak is linear or less than totally branched, their low MW peak is richer in di-block linear species than tri-block species.

These branching calculations generated not only the MWD of the high envelope peak, expressed as $(\overline{M}_{wb})_H$ and $(\overline{M}_{nb})_H$, but also $(\overline{M}_{bp})_H$. The value of $(\overline{M}_p)_H$ was calculated using Equation 16. The ratio of the MW's of the high MW peak to that of the low MW peak, $(\overline{M}_p)_H/(\overline{M}_p)_L$,

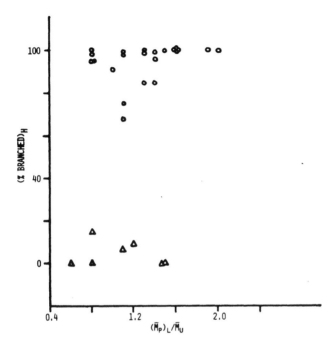

FIGURE 11. $[S(B-DVB)]_x$: Relationship between percentage of high molecular weight copolymer species that is branched and molecular weight of observed low molecular weight copolymer species.

is an expression of the number of linear chains coupled to form the branched chains. It will increase from 1.0 as the amount of branching or coupling occurs. In Figure 12 is plotted $(\overline{M}_p)_H/(\overline{M}_p)_L$ versus $(\overline{M}_p)_L/\overline{M}_u$. The trends suggest, in agreement with those trends in Figure 11, that for those samples whose high MW peak is linear or less than totally branched (triangular data points), their low MW peaks are basically only the initial largely di-block species. Thus, their high MW envelope peak represents species di-coupled to linear tri-block or lightly multi-coupled as branched tri-blocked species. However, for most of the samples, when the high MW peak does represent branched species with high degrees of coupling, branching can occur whether the low MW peak is predominantly linear di-block or tri-block.

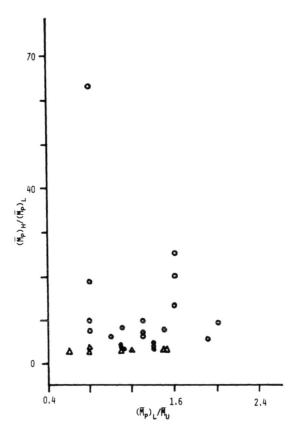

FIGURE 12. [S(B-DVB)]$_x$: Relationship between degree of branching
in copolymer species and molecular weight of observed low molecular
weight copolymer species.

The fact that almost all the samples measured as "100%
Branched" suggests that this scheme for separation of the envelope
into two peaks is a logical separation of branched and linear species.

For each of the samples the MWD and structure of the low and
high MW peaks of the envelope has been determined. This adequately
describes the MWD and structure of the polymer formed during this
rather complex polymerization. This information results from a
mathematical fractionation of the polymer's MWD into distinct peaks

of purer and more uniform composition and then performing certain calculations on these "isolated" peaks. The results indicated the envelope peak to be a blend of a linear species and a branched species. To check these results, the whole polymer was then reanalyzed as if it were a randomly branched copolymer of a certain percentage of branched material. Random branching analyses were performed on the actual experimental GPC peak of the S-B copolymer samples; that is, that envelope peak comprising the two mathematically generated peaks just discussed. The calculated random branching parameters should be appropriate here since presumably the copolymer peak is a mixture of linear and branched species, with the branching concentrated in the higher MW's. Thus, the values of $(\overline{M}_{bp})_E$ and percentage branched in the envelope should have real physical meaning. The branching calculations on the S-B copolymer peak require no prior knowledge of the two mathematically generated peaks. In essence, the envelope peak and the low and high MW peaks can be regarded as representing three different polymers. The branching parameters for the envelope should compare with those of the mathematically generated peaks. This verification should allow further interpretations of the data. In particular, if the calculated percentage branched in the envelope is related to that content of the envelope that is the high MW peak, W_H, this should verify this peak as representing the branched species. In this same fashion, an equality of $\overline{M}_{bp}$ values will serve as verification.

From the values of $[\eta]_E$ and the bi-modal GPC curve of the S-B copolymer species, branching analyses were performed and the results listed in Table 7. Percentage branched ranged from zero to totally branched. The calculated percentage branched correlated well with the percentage of the high MW peak (Fig. 13). For each sample the calculated branching densities $(\overline{M}_{bp})$ for the experimental copolymer peak and the mathematically generated high MW peak also correlated (Fig. 14). These results suggest that the observable high MW peak in the bi-modal copolymer peak is the branched species in the sample. The calculated branching density, $(\overline{M}_{bp})_E$, of the S-B copolymer peak also appears valid.

TABLE 7

Analysis of S-B Copolymer GPC Envelope

Sample	$(\overline{M}_{wb})_E$	$(\overline{M}_{nb})_E$	$(\overline{M}_{wb})_E/(\overline{M}_{nb})_E$	(% Branched)$_E$	$(\overline{M}_{bp})_E$	$(\overline{M}_{bp})_E/\overline{M}_u$
7	736,000	117,000	6.30	79	31,900	0.3
10	20,900,000	136,000	153.9	96	14,600	0.3
13	3,380,000	123,000	27.4	93	17,500	0.3
20	469,000	121,000	3.86	84	28,500	0.5
23	1,020,000	134,000	7.65	97	14,700	0.4
24	6,140,000	166,000	37.0	96	20,500	0.4
28	530,000	102,000	5.18	75	27,600	0.6
29	3,800,000	182,000	20.8	94	24,200	0.4
30	1,440,000	183,000	7.88	91	33,300	0.5
31	350,000	135,000	2.59	52	75,000	1.0
32	562,000	186,000	3.02	68	78,400	0.9
33	349,000	180,000	1.94	13	221,000	2.2
34	278,000	100,000	2.78	4	304,000	3.1
36	164,000	101,000	1.63	0	∞	∞
41	140,000	82,200	1.71	0	∞	∞
42	163,000	79,400	2.05	33	71,200	1.4
43	193,000	105,000	1.84	2	335,000	4.7
47	1,070,000	182,000	5.88	80	52,600	0.4
56	259,000	122,000	2.12	31	110,000	1.5
57	168,000	99,300	1.69	0	∞	∞
60	172,000	102,000	1.68	0	∞	∞
64	189,000	103,000	1.83	3	249,000	2.5
65	664,000	143,000	4.65	69	56,000	0.7
66	460,000	131,000	3.52	78	38,200	0.6
67	314,000	130,000	2.42	58	64,800	0.9
68	197,000	112,000	1.75	0	565,000	6.8
69	226,000	123,000	1.85	3	329,000	3.9
71	228,000	103,000	2.22	2	377,000	3.6

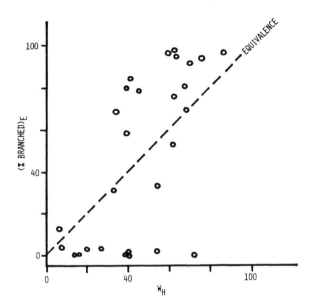

FIGURE 13. $[S(B-DVB)]_x$: Relationship between percentage of copolymer species that is branched and percentage of copolymer species that is a high molecular weight hump.

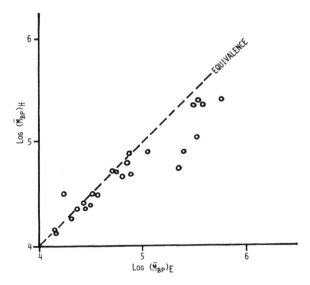

FIGURE 14. $[S(B-DVB)]_x$: Relationship between experimental branching densities determined in different ways.

These various calculations are felt to serve as verification
for the random branching analysis over the entire population of S-B
species. The branching parameters appear to be quantitatively similar to the other, more rigorously generated, parameters. Thus, any
correlations based on branching parameters from the analysis of the
envelope should help in understanding the mechanism of polymerization
and branching. The significance of these branching parameters was
now considered from a different viewpoint. When $(\overline{M}_{bp})_E$ is plotted
versus percentage branched in the envelope (Fig. 15), a smooth curve

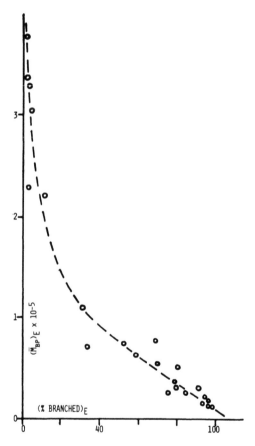

FIGURE 15. $[S(B-DVB)]_x$: Relationship between branching density and
percentage of copolymer species that is branched.

is defined which indicated that as the percentage branched increased, the MW between branch points decreased, that is, branching density increased. The fact that it is a smooth curve also suggests that all these samples are related with respect to their branching structure; perhaps the kinetics of branching is the same at all levels of branching. The MW of the repeat unit, $\overline{M}_u$, did not appear to influence this generalization (Fig. 16). This graph can be interpreted to mean that at high branching percentages and densities, more than one branch point is emanating from the butadiene block length of the repeat unit $(\overline{M}_{bp} < \overline{M}_u)$, while at lower branching percentages and densities, more than one repeat unit is coupled more or less linearly between branch points $(\overline{M}_{bp} > \overline{M}_u)$. The coupling of linear species to form branched species increased smoothly as branching percentage increased (Fig. 17).

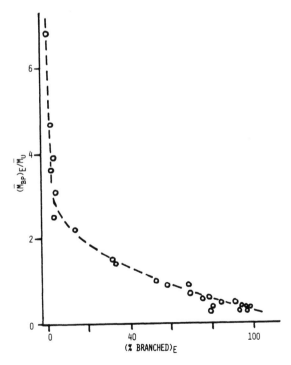

FIGURE 16. $[S(B-DVB)]_x$: Relationship between reduced branching density and percentage of copolymer species that is branched.

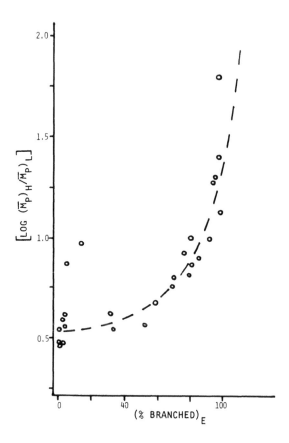

FIGURE 17. $[S(B-DVB)]_x$: Relationship between amount of coupling and percentage of copolymer species that is branched.

These results provide a fairly adequate quantitative representation of the type of polymer being made. Basically, they are in agreement with those conclusions drawn from the earlier more cursory and qualitative evaluation. It appears that the GPC curve provides an accurate description of the mix of branched and linear species. Resolution from the GPC is sufficient to isolate all the different species except for the separation of linear tri-block from linear di-block. These two species differ only by a factor of two in MW and could not be separated from each other. However, partial

separation is achieved and it is good enough to estimate their
contents. The linear species are a mixture of di-block S(B-DVB)
and [S(B-DVB)(DVB-B)S] tri-block. The branched [S(B-DVB)]$_x$ species
varied from light branching to extensive branching. It appears
that, as branching percentage increased, the branching density increased, as did the number of linear units coupled to form the
branched molecule. The interval between branch points can be either
composed of more than one linearly coupled repeat unit (low branching extents) or there can be more than one branch point coming from
each repeat unit (high branching extents).

The results discussed here provide a detailed picture of the
structure of these S(B-DVB) samples with respect to composition, MW
and branching. Although not discussed here, these parameters have
not only been correlated to various physical property measurements
but also combined with spectroscopic and non-polymeric analytical
techniques such as gas chromatography to provide a detailed description of the mechanism of the polymerization of [S(B-DVB)]$_x$ and
[S(I-DVB)]$_x$ block copolymers.

D. Branched Copolymers with
    Abnormal Physical Properties

The previous results were found to correlate well and smoothly to
physical property measurements, such as tensile strength. Tensile
strength was found to increase as the amount of tri-block species
increased. Also, it was relatively straight-forward to assign linear
and branched structures to the observed GPC elution volumes. Referring to Figure 8, there was always an observable peak on the GPC
curve, such as peak B, that correlated well to the structure of a
linear di-block precursor or higher MW tri-block species. For these
polymers the calculated elution volume of the S-B di-block segment
coincides with the peak elution volume of the final block copolymer
or with an elution volume equivalent to a MW lower than the peak MW
of the block (Fig. 18). Based on the GPC curve, the lowest MW major
copolymer fraction has a MW equal to or greater than that calculated

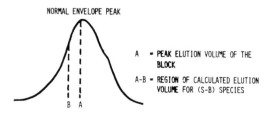

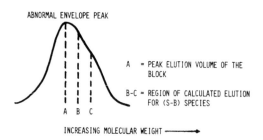

FIGURE 18. $[S(B-DVB)]_x$: GPC curves of two types of copolymer species depicting normal and abnormal branching.

for the basic S-B unit. However, in one set each of $[S(B-DVB)]_x$ and $[S(I-DVB)]_x$ samples, the peak elution volume of the block is greater than that calculated for the S-B unit (Fig. 18). At an initial glance, this would seem to indicate that a great portion of the copolymer is lower in MW than the basic block unit. This is not a likely possibility, knowing the polymerization scheme and since our UV/RI duel detectors indicate that all block species have the same styrene content. The only other explanation is that the basic S-B unit itself is branched and not linear. The elution volume calculation is based on the assumption that the S-B segment is linear in nature. A branched polymer of the same MW as a linear one is smaller in size and will elute at a higher elution volume from the GPC. That is, it will elute as if it were lower MW than it really is. Moreover, it was found that whenever the samples exhibited these "abnormal" GPC curves, they had unusually low tensile strengths. Obviously, the basic S-B units for these copolymers are different in structure. The high tensile blocks are built from segments resembling Structure I, and the DVB branching reaction builds copolymer similar to Structures II and IV where each chain-end has a styrene segment on it. The abnormal low tensile polymers are felt to be built from segments resembling

(VI)

by transfer of the anion to mid-chain DVB pendant vinyl or butadiene double bond, or

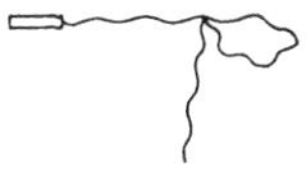

(VII)

if the anion loops back and polymerization back-bites through the same chain's mid-chain DVB pendant vinyl or butadiene double bond. When the DVB couples these chains, there are several loose ends that have no styrene segments on them:

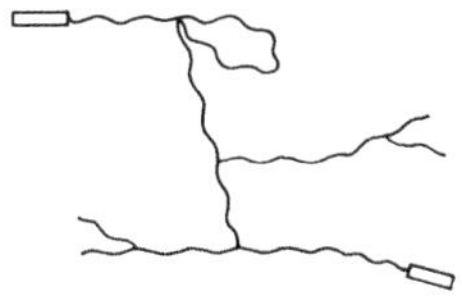

(VIII)

Since these chain ends are not anchored in a styrene segment or physically crosslinked with other butadiene segments, they serve only to detract from the strength of the copolymer.

Some calculations were made on these abnormal samples. Based on the styrene content data and $\overline{M}_s$, the MW of the basic di-block, $\overline{M}_u$, is first calculated, and the elution volume of a linear di-block species of that MW is determined. By comparing the calculated and actual elution volumes, it is possible to calculate the degree of branching. An average of two to three branch points per di-block molecule was found. Considering that the MW of the di-block units

ranged from 50,000 to 70,000 and the (B + DVB) segments was about
70% of that length, incorporating these two to three branch points
within that 35,000 to 55,000 MW length suggests a high degree of
branching, and a sizeable amount of the rubbery block length unan-
chored.

E. Branched Copolymers Containing Vinyl Pyridine

This type of calculation can be applied to other kinds of branching.
Recent studies [14] have shown that during the anionic polymerization of vinyl pyridine (VP) the monomer acts as a tetra-functional
branching agent much like DVB. A series of polymers was made using
a three-step monomer addition scheme

$$Li \xrightarrow{S} S\text{-}Li \xrightarrow{I} S\text{-}I\text{-}Li \xrightarrow{VP} S\text{-}I\text{-}VP \xrightarrow{X} (S\text{-}I\text{-}VP)_x \qquad (18)$$

Conceivably, star-branched block copolymers of functionality x could
be formed. Figure 19 depicts the GPC curves for these polymers,
which indicate that the VP is acting like a crosslinking agent.
Assuming linear species, the high MW peak of the final step is in
many cases ten to fifteen times greater in MW than the original S-I
or S-B unit of the second step. This would indicate the building of
some type of star structure where many S-B or S-I units are connected
through the VP center. Also, the MW of the S-B peak after the third
step reaction did not increase; this peak must not contain much VP
monomer (if any at all). Its linear MW did not increase nor did a
noticeable degree of star-branching develop. UV detector curves
(not shown) indicate that most of the VP is in the high MW peak and
not added to the S-I or S-B unit found in the second step. If a
star-like structure is being formed, then the species are much
greater in MW than the original estimate of 10 to 15 times that of
the S-B or S-I di-block. The actual MW of the species has not been
calculated but it would be possible to do so using a technique similar to that used for branched S(I-DVB) and S(B-DVB) block copolymers.

FIRST STEP S HOMOPOLYSTYRENE

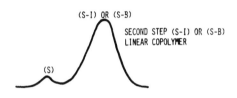

(S-I) OR (S-B)

SECOND STEP (S-I) OR (S-B)
LINEAR COPOLYMER

(S)

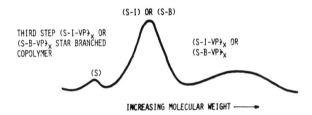

(S-I) OR (S-B)

THIRD STEP (S-I-VP)$_x$ OR
(S-B-VP)$_x$ STAR BRANCHED
COPOLYMER

(S-I-VP)$_x$ OR
(S-B-VP)$_x$

(S)

INCREASING MOLECULAR WEIGHT ⟶

FIGURE 19. [S(B-VP)]$_x$: GPC curves of star-branched tri-block
copolymer.

F. Analysis of Other Types of Block Copolymers

Two samples of [S(I-DVB)]$_x$ were compared to commercial samples of
Kraton 1107 and Solprene 418, which are reportedly prepared in completely different ways and are suspected to have radically different
configurations. All four samples contained varying amounts of
homopolystyrene (Fig. 20-23). The low MW peak was clearly separated
from the high MW envelope of the copolymer species, which appeared
as either bi-modal or a skewed uni-modal peak with distinct shoulders.
Table 8 lists homopolystyrene content, MW, and intrinsic viscosity,
copolymer compositions, intrinsic viscosities of the sample and envelope, and the MW of the theoretical di-block repeat unit. Based on
the assumed mode of polymerization, several specific calculations
were made as done for the [S(B-DVB)]$_x$ copolymers.

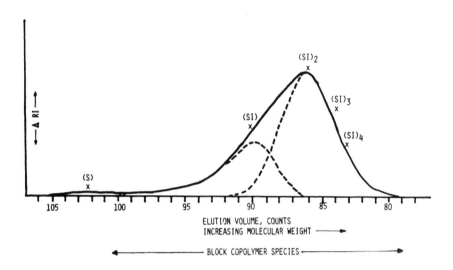

FIGURE 20. Kraton 1107: Assignment of polymer structures within its GPC curve.

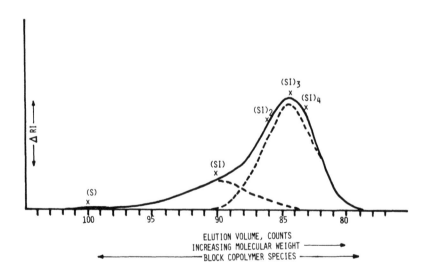

FIGURE 21. Solprene 418: Assignment of polymer structures within its GPC curve.

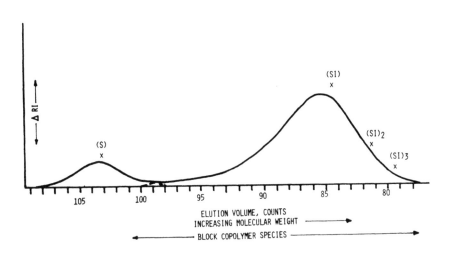

FIGURE 22. [S(I-DVB)]$_x$ Sample I-129: Assignment of polymer structures within its GPC curve.

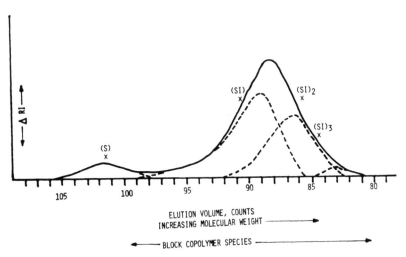

FIGURE 23. [S(I-DVB)]$_x$ Sample I-134: Assignment of polymer structures within its GPC curve.

TABLE 8

Analysis of Commercial Block Copolymers

	Kraton 1107	Solprene 418	Sample I-129	Sample I-134
(% Styrene)$_{total}$	13.7	15.1	12.8	17.6
W_s	1.0	0.7	9.0	5.7
(% Styrene)$_E$	12.8	14.5	4.4	14.3
$[\eta]_t$	1.296	1.473	1.752	1.099
$[\eta]_s$	0.116	0.124	0.106	0.121
$\overline{M}_s$	10,700	12,100	8,900	11,500
$\overline{M}_u$	83,500	83,400	202,000	80,400
$[\eta]_E$	1.308	1.482	1.915	1.158
W_H	68	72	100	38 + 4[a]
W_L	32	28	0	58

[a]Two high MW peaks; the higher of these two peaks was smaller in relative area.

The patent literature suggests the following polymerization mechanism as the basis of Kraton 1107's structure:

$$\text{Li} \xrightarrow{S} \text{S-Li} \xrightarrow{I} \text{S-I-Li} \xrightarrow{f=2} \text{S} - \text{I-I} - \text{S}. \tag{19}$$

Assuming this mechanism to apply, the GPC curve may contain only up to three distinct species: S, S-I (MW = $\overline{M}_u$), and S—I-I—S or (S-I)$_2$, exactly twice the MW of S-I due to the linear di-functional coupling reaction. The MW's of these species, and higher MW branched species, are multiples of $\overline{M}_u$. Their expected GPC peak elution volumes are shown in Figure 20. The envelope peak was deconvoluted into the two Gaussian-shaped peaks shown. Their peak elution volumes were found to coincide to where the indicated species would be expected to elute. It is difficult to envision species of (S-I)$_3$ or (S-I)$_4$ being predominant components of the envelope peak.

The patent literature suggests the following polymerization mechanism for Solprene 418:

$$\text{Li} \xrightarrow{S} \text{S-Li} \xrightarrow{I} \text{S-I-Li} \xrightarrow{f} (S-I)_f. \tag{20}$$

Assuming this mechanism to apply, the GPC curve may contain only up to three distinct species: S, S-I and $(S-I)_f$, where f is the functionality (probably 2, 3 or 4) of the coupling agent. When f = 3 or 4, $(S-I)_f$ represents a star branched tri-block species. The MW's of the theoretically possible species again are multiples of $\overline{M}_u$. Their expected peak elution volumes are shown in Figure 21. It is evident from the GPC curve that it is probably a blend of S-I and tri-functional star-branched $(S-I)_3$. It is difficult to envision the GPC curve as being comprised of either linear tri-block $(S-I)_2$ or tetra-functional star branched $(S-I)_4$. The GPC curve was separated into the two peaks shown in Figure 21. The low MW envelope peak was forced to coincide to that of the elution volume of (S-I), while the high MW envelope peak was allowed to conform to that of the envelope.

For the two experimental samples I-129 and I-134 the polymerization mechanism that was described earlier was assumed

$$Li \xrightarrow{S} S\text{-}Li \xrightarrow{I+DVB} S(I\text{-}DVB)Li \xrightarrow{x} [S(I\text{-}DVB)]_x \qquad (21)$$

The GPC curve may contain homopolystyrene S, linear di-block S(I-DVB), linear tri-block $S(I\text{-}DVB)_2$, and randomly branched tri-block $[S(I\text{-}DVB)]_x$, where $x \geqslant 3$. The MW's of these entities are multiples of $\overline{M}_u$. Elution volumes of a few homologues are marked in Figures 22 and 23.

Specific branching analyses were then made, placing different constraints on each sample, depending on its polymerization mechanism and assumed branching model.

For Kraton 1107, the calculated value of $[\eta]_E$ was used to evaluate the presence of branching in the envelope. The results, listed in Table 9, indicated Kraton 1107 to be linear polymer. Furthermore, the calculated MW's of the envelope peak also suggested only di-block and tri-block linear species to be the predominant species and not higher MW branched homologues. The average degree of coupling calculated from the branching analysis, 1.5, is in satisfactory agreement with that calculated from the GPC curve, 1.52, using

TABLE 9

Analysis of Commercial Block Copolymers

	Kraton 1107	Solprene 418	Sample I-129	Sample I-134
$(\bar{M}_{wb})_E$	156,000		200,000	119,000
$(\bar{M}_{nb})_E$	104,000		117,000	89,900
$(\bar{M}_{wb})_E/(\bar{M}_{nb})_E$	1.50		1.71	1.32
$(\bar{M}_p)_E$	127,000		153,000	103,000
(% Branched)$_E$	0		0	0
$(\bar{M}_{bp})_E$	∞		∞	∞
$(\bar{M}_p)_E/\bar{M}_u$	1.5		0.8	1.3
$(\bar{M}_w)_L$	74,500	91,300		85,100
$(\bar{M}_n)_L$	59,300	70,200		71,500
$(\bar{M}_p)_L$	66,500	80,000		78,000
$(\bar{M}_w)_L/(\bar{M}_n)_L$	1.26	1.30		1.19
$(\bar{M}_p)_L/\bar{M}_u$	0.8	1.0		1.0
$(\bar{M}_w)_H$	194,000	260,000		151,000 (≥ 378,000)
$(\bar{M}_n)_H$	159,000	207,000		133,000 (≥ 363,000)
$(\bar{M}_p)_H$	176,000	232,000		142,000 (≥ 370,000)
$(\bar{M}_w)_H/(\bar{M}_n)_H$	1.22	1.26		1.14 (≥ 1.04)
$(\bar{M}_p)_H/\bar{M}_u$	2.1	2.8		1.8 (≥ 4.6)

$$\frac{1}{f} = \frac{W_H}{100 \cdot f_H} + \frac{W_L}{100 \cdot f_L} \tag{22}$$

with $f_H = 2$, $f_L = 1$. No higher MW branched homologues are present
in the sample, since no peaks or shoulders were found in the necessary elution volume regions. Calculated MW's of the two deconvoluted
peaks are also in good agreement with those predicted for these species.

For Solprene 418, the two deconvoluted peaks were analyzed
separately. For the low MW peak, the result suggests linear di-
block. The results of the high MW peak suggests it to be a tri-block

3-legged star-branched block copolymer. Roughly 2/3 of the block
species is star-branched tri-block, 1/3 is linear di-block. Little
evidence for either linear tri-block or higher functionality branched
tri-block is seen.

For the two samples I-129 and I-134, it was not that evident
exactly how to deconvolute the envelope peak. Therefore, random
branching analyses were done on the envelope. Both envelopes were
found to be linear polymer. As evidenced by their values of
$(\overline{M}_p)_E/\overline{M}_u$, sample I-129 was predominantly linear di-block repeat unit,
and sample I-134 was mainly (approx. 60%) linear di-block with some
linear tri-block (approx. 40%) species present. The GPC curve of
sample I-134 could be easily deconvoluted into two major Gaussian
curves eluting close to those of these species, with perhaps a small
fraction of higher MW branched species. The calculated degree of
coupling using equation (22) was 1.28, in good agreement with the
earlier calculated value. MW results on each peak were in good
agreement with the assumed species. In parenthesis in Table 9 are
the estimated results for the highest MW deconvoluted peak.

It was found for all four samples, when considering the MWD of
the coupled peaks, that they were always narrower than the MWD of
the bi-block precursors. However, in all cases the narrowing of the
MWD upon coupling was not as severe as that predicted by Kraus [15]
for coupling of polydisperse anionic polymers.

III. CONCLUSIONS

The versatility of this method for determining the structure of
branched block copolymers has been demonstrated. All that is required
is a GPC curve having sufficient resolution for the various possible
species, and knowledge of the total composition and intrinsic viscosity for the sample. By knowing or assuming a branching reaction,
total structural configuration of the polymer can be quantitatively
made, including presence and amount of linear first block homopolymer,
linear di-block, linear tri-block and branched tri-block homologues,

and their MW's and branching extents. Several examples have been shown for the analysis of block copolymers of widely differing compositions, MW's, and types and extents of branching. By themselves, these results do not provide a definitive proof of structure, but they can be verified by other more absolute MW and branching measurements. Results generated by this method are believed to be quite reliable and they will serve as useful estimates to guide programs of synthesis and optimization of properties.

ACKNOWLEDGMENTS

The author wishes to acknowledge the assistance of Dr. J. L. Corey for the spectroscopic analyses, Mr. D. P. Cardina for the GPC, intrinsic viscosity and gel analyses, and Mr. C. D. Shuster for the MW and branching calculations. The author also thanks the Goodyear Tire and Rubber Co. for permission to publish this work.

REFERENCES

1. M. R. Ambler, J. Appl. Polym. Sci., 21, 1655 (1977).

2. M. R. Ambler, R. D. Mate, and J. R. Purdon, Jr., J. Polym. Sci., Polym. Chem. Ed., 12, 1771 (1974).

3. M. R. Ambler, R. D. Mate, and J. R. Purdon, Jr., J. Polym. Sci., Polym. Chem. Ed., 12, 1759 (1974).

4. M. R. Ambler and D. McIntyre, J. Polym. Sci., Polym. Letters Ed., 13, 589 (1975).

5. Polymer Handbook (J. Brandrup and E. H. Immergut, eds.), Interscience, New York, 1967.

6. J. N. Anderson, M. L. Barzan, and H. E. Adams, Rubber. Chem. Techn., 45(5), 1270 (1972).

7. F. S. C. Chang, in Polymer Molecular Weight Methods, Advances in Chemistry Series 125 (M. Ezrin, ed.), American Chemical Society, Washington, D. C., 1973, p. 154.

8. L. H. Tung, J. Appl. Polym. Sci., 24, 953 (1979).

9. C. Kuo, Doctoral Thesis, Univ. of Akron, Akron, Ohio, 1973;
    M. R. Ambler, Doctoral Thesis, Univ. of Akron, Akron, Ohio,
    1975.

10. T. Tanaka, T. Kotaka, K. Ban, M. Hattori, and H. Inagaki,
    Macromolecules, 10(5), 960 (1977).

11. H. E. Adams, in Gel Permeation Chromatography (K. H. Altgelt
    and L. Segal, eds.), Dekker, New York, 1971, p. 391.

12. M. R. Ambler, J. Appl. Polym. Sci., 20, 2259 (1976).

13. R. T. Prudence, U.S. Pat. 3,949,020, April 6, 1976.

14. A. R. Luxton, A. Quig, M. J. Delvaux, and L. J. Fetters,
    Polymer, 19(11), 1320 (1978).

15. G. Kraus and J. T. Gruver, J. Polym. Sci., Part A, 3, 105
    (1965).

POLYSTYRENE BONDED SILICA AS GPC PACKING:
A VARIABLE PORE DIAMETER PACKING CONCEPT IN GPC

Benjamin Monrabal*

Chemistry Department
Virginia Polytechnic Institute
Blacksburg, Virginia

SUMMARY

A technique to chemically bond a vinyl polymer onto the silica surface was developed. Noncrosslinked polymer layers of various thicknesses were attached to silicas of various mean pore sizes. When the reaction is carried out in the presence of divinylbenzene, a new porous network is formed inside the silica pores. The polystyrene bonded silica particles possess a noticeable flexibility and can be used in various chromatographic modes.

A new concept of a variable pore diameter packing for size exclusion chromatography was described. The polymer bonded layer can be swollen to a different extent with mobile phases of different solvation power; thus, the pore size distribution can be shifted. However, secondary mechanisms interfere with the size exclusion process when mobile phases having solubility parameters very different from the polymer layer and solute are used.

The attachment of a polystyrene layer to the walls of a narrow pore size distribution silica packing was shown to produce a shift in the molecular weight range of separation by size exclusion with a net improvement in selectivity for the new range.

I. INTRODUCTION

Since the development of GPC by Moore in 1964 [1], and until very recently, polystyrene beads have been the only available GPC packing for the analysis of organic polymers, and today they are probably

*Current affiliation: Dow Chemical Iberia, Tarragona, Spain.

still the most popular packing. With improved mechanical stability
and smaller particle size than the initial Styragel packing, the
more recent polystyrene beads [2-4] are being used quite successfully
in high performance GPC; nevertheless, packed columns of polymer
beads, although highly crosslinked, still deserve some special care
in their handling due to the nontotal rigidity and swelling characteristics of the organic polymer matrix.

Silica particles, on the other hand, with higher mechanical
stability and available in various pore sizes are gaining wide
acceptance in recent years as high efficienty packings for GPC [5-7],
especially in the biochemistry field [8-11], where the current gel
filtration packings lack the rigidity to withstand the high pressures
of modern HPLC. Silica, however, with a heterogeneous surface activity suffers from adsorption of polar molecules which invalidates the
molecular weight scale. A great deal of attention has been given to
the elimination of the adsorption problems of biomolecules in aqueous
GPC and a lot of success has been achieved in recent years [8-14].
In addition to their surface activity and pH limitations, silica
particles also suffer from another disadvantage in that they are
difficult to produce, with good size exclusion characteristics, at
the lower mean pore diameters [5]. Thus, silica packings lack in
separation power for low molecular weight compounds.

It is obvious that the combination of the rigid structure of
a silica matrix and the homogeneous and low activity of an organic
polymer surface, such as polystyrene, would be desirable for a high
performance GPC packing. A polymer multilayer attached to the
silica walls would deactivate the surface to a greater extent than
the classical "monolayer;" however, the difficulties in achieving a
thin and uniform layer of polymer on a porous silica surface are
obvious [5]. Thick layers of polymer attached on the surface could
produce a deleterious high resistance to mass transfer [15-17].

The first part of this research deals with an original approach
to chemically bond a polystyrene layer into the walls of a totally
porous silica. The reaction steps are similar to those used by
Caude and Rosset [18].

The polystyrene bonded silica particles possess a noticeable flexibility and can be used in the various chromatographic modes such as GC [19], IEC, Reverse Phase HPLC, Phase Distribution Chromatography [20-21], and GPC, and when the polymerization conditions are such so as to produce a porous organic structure within the silica pores, the large surface areas achieved together with the high rigidity of the particles may be useful for other applications such as solid phase peptide synthesis and supported catalysts. Also, it can be expected that carrying the polymerization with acrylamide instead of styrene monomer, a polyamide bonded silica with good properties for aqueous GPC separation of proteins could be achieved.

Partial characterization of the synthesized particles is reported as well as some preliminary data on the degree of surface deactivation achieved. The effect of the polymer layer thickness on the molecular weight separation range is also studied.

The second part of this paper deals with the description of a new variable pore diameter concept in GPC which is applied to the polystyrene bonded silica particles. The principle is to vary the pore diameter by swelling the bonded layer to a different extent through the use of solvent mixtures of various polarities. The limitations of this concept and the difficulties found in their application are discussed.

II. EXPERIMENTAL

A. Apparatus

Polymerization was carried out in a stainless steel tubular reactor with different lengths (15-45 cm) and 1/2 in. or 3/8 in. O.D., depending on the amount of silica to be treated. A Milton Roy (LDC, Riviera Beach, Fla.) pump model DB1-29R with a capacity of 124 ml/hr and a maximum pressure of 3000 psi was used to pump the monomer solution through the reactor.

A Bendix gas chromatograph model 2110 served as the reactor oven and was modified to adapt a switching valve to the reactor and an auxiliary gas line (Fig. 1). The chromatograph itself was used to

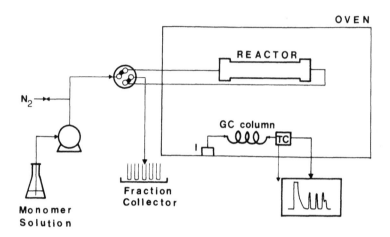

FIGURE 1.   Polymerization apparatus.

monitor the reaction by analyzing samples collected at the reactor
output.   A column 6 ft. long 1/8 in. O.D. of 4% OV-17 on Chromosorb
G 80/100 mesh was used at the reactor temperature to analyze styrene
and divinylbenzene isomers.

An LC unit was assembled from the following components: a
Milton Roy minipump (240 ml/hr and 1000 psi) with a pulse dampener
LDC 709, a Spectra-Physics column oven model 748, a Valco injection
valve, a Waters Associates refractive index detector model R-401 and
a Waters Associates syphon counter installed inside the oven.

B.   Reagents and Materials

Styrene (99%) and benzoyl peroxide were obtained from Aldrich Chemical So. (Milwaukee, Wis.).   Divinylbenzene, practical grade (39%
m-DVB, 15% p-DVB, 43% ethylvinylbenzene) was obtained from Pfaltz
and Bauer (Stamford, Conn.).   The various silane reagents used were
obtained from Petrarch Systems Inc. (Levittown, Penn.)   Polystyrene
standards were obtained from Waters Associates (Milford, Mass.) and
from ArRo Laboratories (Joliet, Ill.).

Lichrospher silica supports SI-100, SI-500 and SI-1000, in 10
μm particle size, were donated by E. M. Laboratories (Cincinnati,
Ohio).   Porasil A and F were obtained from Analabs (North Haven, Ct.).

C. Silylation

Large batches of silica, either Porasil or Lichrospher of various mean pore diameters, were treated with vinyl silane reagents. The pretreatment and reaction conditions for the Porasil samples have already been described [19]. Lichrospher samples were heated at 220°C for 15 min under nitrogen purge prior to reaction with the silane reagents.

Typically 50 g of silica were mixed with 100 ml of pure vinyldimethylethoxysilane and refluxed for 20 hours. Silanol coverages of ca. 50% were obtained.

D. Polymerization

Polymer bonded silica is achieved through free radical polymerization by first attaching a vinyl group onto the silica surface. The vinyl group serves as the linkage point from which the polymer chain will start growing when reacted with the various monomers. If the

$$
\begin{array}{c}
CH=CH_2 \\
| \\
CH_3-Si-CH_3 \\
| \\
O \\
| \\
\text{---}Si\text{---} \\
/ | \backslash
\end{array}
\quad + \quad
\begin{array}{c}
CH=CH_2 \\
| \\
\bigcirc
\end{array}
\xrightarrow[100°C]{toluene}
\begin{array}{c}
\bigcirc \quad \bigcirc \\
[-CH-CH_2-CH-CH_2-CH-CH_2-] \\
| \\
CH_3-Si-CH_3 \\
| \\
O \\
| \\
\text{---}Si\text{---} \\
/ | \backslash
\end{array}
$$

reaction solution contains only monovinyl monomer, linear polymer-chains will be formed hanging from the silica surface. When the reaction solution contains divinyl monomers, a crosslinked network will be formed which may result in a new organic porous structure inside the pores of the silica skeleton.

The polymerization is carried out in a tubular reactor, with no recycle, and at reaction conditions which result in slow chain growth. First, the reactor is filled with the vinyl modified silica particles. Then, the polymerization begins by continuously pumping a monomer and initiator solution through the reactor which is kept at 100°C. Typically a solution of 40 g of styrene, 1 g of benzoyl peroxide and up to 800 ml of toluene is pumped at 1 ml/min for a few hours.

The whole process of chain growth and termination is known to take place in solution in a relatively short time [22]; thus, polymer chains will be initiated and terminated along the reactor length. Eventually, a steady state will be achieved in solution with the rate of initiation being equal to the rate of termination at each particular point of the reactor. On the other hand, the free vinyl groups and the active chain radicals attached to the silica surface will decrease with time and eventually disappear by termination; therefore, no further linear propagation will take place on the polymer bonded to the silica. However, given the peculiar characteristics of this process in which the bonded polymer is kept in continuous contact with a high and constant concentration of radicals and, based on the experimental results, reinitiation by chain transfer of growing polymer or initiator to polymer may take place [23–25] resulting in a continuous growth of the bonded chains with a branched structure.

The geometry of a tubular reactor allows the continuous elimination of the polymer formed in solution. This reactor, however, does suffer per se from a variable reagent and product concentration along its path. To avoid a nonuniform attachment through the reactor length, monomer conversion is adjusted to around 10% by varying the flow or temperature, and the flow direction is switched every hour.

Once the reaction is considered complete according to previous reactions and the volume pumped, the reagent solution is replaced by pure solvent and pumping continues for 2 hours to eliminate the nonbonded polymer from the reactor bed.

The main characteristic of this technique is the fact that the thickness of the bonded polymer layer increases steadily with time while reaction conditions are such as to produce low molecular weight chains in solution to avoid clogging the silica pores or precipitating between particles when divinyl monomers are used. It can be concluded that the use of toluene as carrier which has a relatively high transfer constant, together with the dilute solutions used, the large concentration of initiator, the low conversion and the short residence time, may explain the good control achieved in the reactions.

Finally, the proper considerations should be given to the hazard aspects involved in the use of peroxides and in the reaction itself.

III. RESULTS AND DISCUSSION

A. Type of Attachment

The polymer that remained in the silica particles, after thorough washing, was measured by elemental analysis, and it was considered as attached polymer. However, adsorption forces of macromolecules can be large [26], and irreversible physical trapping or entanglement of polymer molecules growing in a highly porous skeleton can take place [27]. In order to determine whether chemical or physical attachment has occurred, several experiments were planned using both modified (vinyl bonded on the surface) and unmodified silica (Lichrospher SI-100). The attachment yield which is defined as the fraction of polymer bonded over the total polymer produced is used for the comparisons. Results are shown in Table 1. When the reaction is carried out with styrene monomer the largest part of the polymer is chemically bonded as shown by the higher attachment yield of sample 1 versus sample 2. However, when divinylbenzene is used the physical attachment appears to be, as expected, very important.

B. Characterization

A large number of reactions were carried out on Porasil A and F with both styrene (STY) and divinyl benzene (DVB). Table 2 shows some of the properties measured on these samples.

TABLE 1

Polymer Attachment Lichrospher SI-100

Sample No.	Vinyl bonded	Monomer	Monomer reacted per g silica	Polymer bonded (%)	Attachment yield (%)
1	YES	Styrene	0.5	7.0	14.0
2	NO	Styrene	0.94	5.4	5.8
3	NO	DVB	0.28	10.6	38.0

TABLE 2

Polystyrene Bonded Silica Characterization

#	Packing	Organic matter content (%)	Pore volume (%)	C/Si	HETP (mm)	Surface area (m^2/g)
4	PORASIL A	0	2.1	0.31	0.8	362
11	A-STY-1	12.0	--	0.90	1.1	306
13	A-STY-3	35.0	0.95	--	14.0	96
14	A-DVB-1	29.0	--	9.8	1.9	400
15	A-DVB-2	31.0	--	--	3.0	695
5	PORASIL F	0	2.0	0.47	0.9	13
7	F-TM	<0.5	--	0.79	--	--
18	F-STY-1	0.6	1.9	--	--	--
19	F-STY-2	1.0	--	2.1	1.2	--
20	F-STY-3	3.0	1.8	--	1.3	13
21	F-STY-4	5.7	1.6	11.4	2.5	--
23	F-DVB-2	5.7	--	5.6	1.7	--
22	F-DVB-1	35.0	0.3	∞	10.0	208

Note: The pore volume was measured by GPC in a standard column using polystyrene Mw = 800 for Porasil and n-hexane for porasil A packings.

There is an acceptable correlation between the amounts of polymer bonded and the pore volumes measured by GPC which is an indication of the good control in the polymerization technique.

The carbon to silicon ratios (column 3), measured by ESCA, give a rough idea of the degree of surface coverage and deactivation achieved. A trimethyl silane treatment (packing 7 F-TM) results in lower C/Si ratio than those of polystyrene bonded silica (packings 19, 21, 22 and 23). However, large polymer attachments may result in a high resistance to mass transfer as it is shown by the large HETP values (column 4) measured by GC of packings 13 and 22 compared to those of the starting silica material (packings 4 and 5).

Surface deactivation was also related to the behavior of small polar molecules in GC. The peak shapes in Figure 2 show that the deactivation obtained by trimetyl or diphenyl silane treatment (packings 6 and 8) is further improved by polymer bonding on Porasil A. Alcohols, however, are eluted with large tailing. A more complete deactivation was achieved on Porasil F as it could be expected from their larger pores and lower surfaces areas [19].

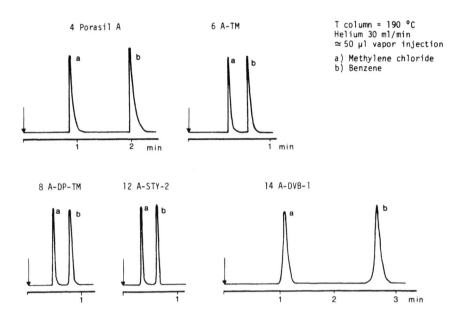

FIGURE 2. Chromatographic behavior of Porasil A and chemically modified Porasil A. TM = trimethyl; DP = diphenyl; STY = styrene monomer; DVB = divinylbenzene monomer.

The surface areas, measured by nitrogen adsorption (Table 2), show that when DVB is reacted an organic porous structure is formed; thus, the surface area of Porasil F increased from 13 m^2/g up to 208 m^2/g (packing 22). However, when only styrene is reacted the surface area is reduced as it should be expected from the reduction of the silica pore diameter.

The differences between the STY and DVB polymer bonded silicas were also studied by the heats of interaction with small molecules. The results have already been reported [19].

IV. VARIABLE PORE DIAMETER PACKING CONCEPT

A good GPC packing in terms of the skeleton structure should have a narrow pore size distribution (PSD) and a high pore volume to achieve the highest selectivity and peak capacity.

The interest in narrow PSD packings, however, requires the existence of various mean pore size packings. On GPC, capacity and selectivity cannot be expanded, as in other chromatographic modes, by temperature programming or solvent gradient due to basic mechanism and limited pore volume.

In 1970, Giddings [28] proposed a method for expanding the capacity (abscissa axis in the calibration curve) and improving resolution in GPC, which he refers as Programmed Exclusion Chromatography. This method, however, has not been further developed. In this research, a new concept is proposed to shift the molecular weight range of separation (ordinate axis in the calibration curve) for a particular packing. The principle is to chemically bond a polymer layer to the surface of a narrow PSD silica and, through the use of solvent mixtures of various polarities, vary the pore diameter by the different swelling of the bonded layer. Referring to Figure 3, the PSD curve

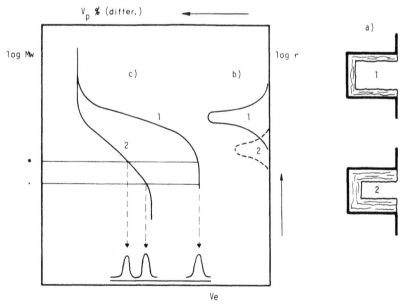

FIGURE 3. Schematic diagram of the effects produced by varying the bonded polymer layer thickness. (1) Thin or shrunken layer; (2) thick or swollen layer; (a) pores (cylindrical); (b) PSD curves; (c) calibration curves.

1b corresponds to the layer thickness shown in 1a and results in
calibration curve 1c. This layer thickness would correspond to a low
swelling power of the solvent; thus, the pore diameter would be the
maximum. If we increase the percentage of the solvent which inter-
acts more strongly with the polymer layer, a decrease in pore diameter will result as shown in 2a. This will produce a shift in the
PSD (2b) and a corresponding change in the calibration curve (2c).
The calibration curve 2c has shifted the separation range to lower
molecular weights. However, the total pore volume decreases as the
polymer layer swells counteracting the improvement of selectivity.

It would be interesting to know at which broadness of the PSD
the deleterious effect of the pore volume decrease will overcome
the improvement obtained by the PSD shift. To better understand
this concept a simplistic model was prepared assuming that the PSD
follows a gaussian distribution:

$$V_p = \frac{1}{2\pi} \cdot e^{-\frac{(r-r_m)^2}{2\sigma^2}} \tag{1}$$

where V_p is the pore volume, r is the pore radius, r_m is the most
probable pore radius and σ is the standard deviation of the PSD.

The new pore volume after decreasing the radius by a thickness
t will be given by

$$V_p' = \frac{1}{2\pi} \cdot e^{-\frac{(r-r_m)^2}{2\sigma^2}} \cdot \frac{(r-t)^2}{r^2} \tag{2}$$

which will correspond to the new radius r - t. In Figure 4, both
the original and new PSD are plotted for various standard deviations assuming that the thickness t (10 Å) is the same for all the
pores. As the PSD is broadened, the net improvement in pore volume
for the smaller radii, as a result of the PSD shift, is reduced and
finally overcome by the decrease in the magnitude of the pore volume.

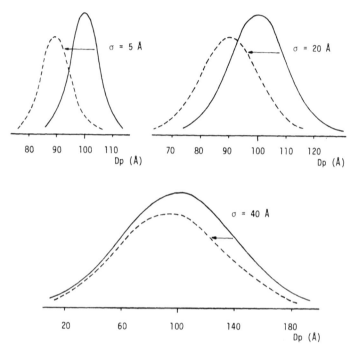

FIGURE 4.   Theoretical shift of the pore size distribution curve
(differential) upon attachment of a polymer layer (10 Å) to the walls
of silica particles.   Dp = 100 Å.

    With the current GPC packings, having a fairly sharp PSD, a net
improvement would be expected (assuming that the bonded layer were
uniformly attached).   Also, if the macromolecules to be analyzed have
a similar solubility parameter to that of the bonded layer, they will
swell and their size will change in a similar manner; this results in
an "extra" shift in the molecular weight range.   Thus, a lower increase in layer thickness would be required (less pore volume loss)
to cover a given molecular weight range.

A.   Reduction of Silica Mean Pore Diameter

Various packings have been prepared to prove that by bonding a polymer
layer to the silica surface, the resulting reduction of the pore diameter shifts the calibration curve in such a way as to increase the
selectivity in the lower molecular size range.

All the GPC data were obtained using toluene as the mobile
phase at 1 ml/min and a column temperature of 40°C. Columns were
all 45 cm length and 7.1 mm I.D.. The calibration curve was first
determined on the starting silica material using narrow polystyrene
standards, and then on the polystyrene bonded silica.

Figure 5 shows the calibration curves of Lichrospher SI-500
unmodified (packing 4) and with polystyrene attached containing 3%
of organic matter (packing 4B). The expected shift in the calibration curve was obtained. The selectivity has decreased at the high
molecular size range, but it has improved at the low molecular size.
This can be shown in Figure 5 by the change in the slope and by the
difference in elution volume corresponding to two solute sizes. The
analysis of low molecular weight hydrocarbons also showed the improvement; the separation of C_{30} and C_6 is better in packing 4B than
in the original silica, packing 4.

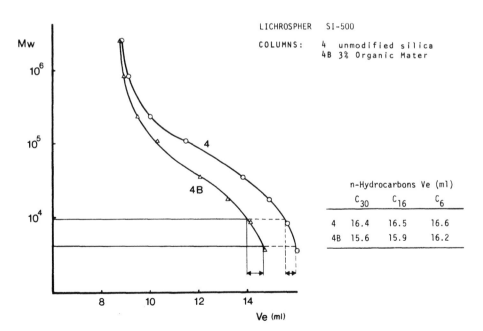

FIGURE 5. SEC calibration curve shift upon attachment of a polymer
layer to the silica surface. Columns 3A and 3C.

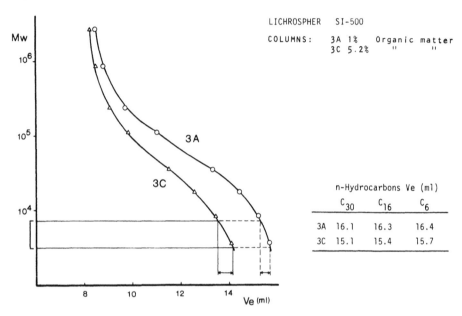

FIGURE 6. SEC calibration curve shift upon attachment of a polymer layer to the silica surface. Columns 3A and 3C.

The same experiments were carried out with packing 3A, which corresponds to the vinyl modification of Lichrospher SI-500, and packing 3C which corresponds to the polymer bonded silica containing 5.2% of organic matter (Fig. 6). Similar results as above were obtained with an improvement in selectivity at the low molecular size range.

B. Variable Pore Diameter Packing: Results and
   Discussion

In the previous section the GPC data were obtained with toluene as the mobile phase. Toluene has a solubility parameter (δ = 8.9) very similar to that of polystyrene (δ = 9.1); therefore, the polymer layer was expected to be the maximum swollen state and the minimum pore diameter would have resulted. At this point, it was interesting to know whether the layer thickness could also be modified by a nonswelling solvent mobile phase; thus, the variable pore diameter concept could be achieved by using as the mobile phase mixtures of toluene with various percentages of the nonswelling solvent. This solvent should have a different solubility parameter than polystyrene, but still should dissolve the polystyrene standards used in the calibration and it should not contribute in the separation process to any secondary mechanisms such as adsorption or partition. Probably there is not such a solvent for the system being experimented within this research, and a compromise has to be accepted.

C. DMF-Toluene Mixtures as the Mobile Phase

Dimethylformamide (DMF), with a solubility parameter of δ = 12.1, appears to be the most appropriate choice when considering a solvent more polar than polystyrene. However, it has been shown that when DMF is used as a mobile phase in GPC with a polystyrene packing (Styragel), the polystyrene standards are delayed. This is explained in terms of adsorption [29,30] and partition-adsorption [31] effects which are added to the exclusion mechanism. Nevertheless, DMF was chosen expecting that mixed in different proportions with toluene, the secondary effects could be avoided while the shift in the pore diameter would still occur.

Figure 7 shows the calibration curves obtained using various mixtures of toluene (TOL) and DMF as the mobile phase at 40°C and 1 ml/min. The difference between the curves corresponding to pure toluene and pure DMF should be explained in terms of (1) adsorption or partition and (2) because of the expected change in the pore diameter. Using a mobile phase TOL 10%, DMF 90% shifts the calibration curve as expected although it is not possible to elucidate which mechanism is more important, (1) or (2).

Surprisingly, when a TOL 50%, DMF 50% mobile phase is used, the calibration curve is shifted towards the lower elution volumes passing that of the pure toluene, and with a solvent TOL 90%, DMF 10% the calibration curve achieves the shortest elution volumes. As the DMF percentage is further decreased, the calibration curve approaches that of pure toluene.

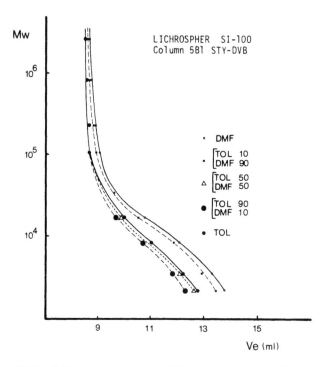

FIGURE 7.   SEC calibration curve shift with toluene-dimethylformamide
mixtures as the mobile phase.

A similar phenomenon, but to a smaller degree, was observed when
these experiments are carried out on unmodified silica.   Thus, it
appears that the cause of this phenomenon is a solvent solute interaction.   It seems to indicate that the polystyrene would swell to a
higher extent in a mixture of toluene and DMF than in pure toluene.
Or in other words, the more polar environment (TOL 90%, DMF 10%)
would produce a driving force to increase the association of the
solute-toluene pair.

Table 3 shows the elution volumes of n-hydrocarbons using the
same solvent mixtures as in Figure 7.   The minimum values also correspond to the solvent mixture TOL 90%, DMF 10%.   Moreover, an inversion
of the elution volumes results as the percentage of DMF is increased.
This is the result of partition-adsorption taking place and overcom-
ing the exclusion mechanism.

TABLE 3

Hydrocarbon Elution Volumes on Polystyrene
Bonded Silica (Column 5B1 SI-100)[a]

| Mobile | Volume | Elution volumes (ml) | | |
Phase	percentage	C_{30}	C_{16}	C_6
Toluene	100	14.0	14.5	15.3
Toluene DMF[b]	95 5	13.6	14.2	15.0
Toluene DMF[b]	90 10	13.6	14.2	14.9
Toluene DMF[b]	50 50	14.4	14.8	15.2
Toluene DMF[b]	10 90	16.2	16.3	16.2
DMF[b]	100	16.6	16.5	16.4

[a]Column 5B1 contains 15% organic matter. Experiments carried out at 1 ml/min and 40°C.

[b]DMF = N, N dimethylformamide.

D. Cyclohexane-Toluene Mixtures as the Mobile Phase

Thus far the use of toluene-DMF mixtures as the mobile phase did not result in a successful exclusion process in which the packing pore diameter could be varied. Therefore, it was decided to repeat the experiments with mixtures of toluene and another nonswelling solvent, this time the solvent being less polar than polystyrene. Cyclohexane (CHX) appeared to be an appropriate choice. With a solubility parameter (δ = 8.2) lower than that of polystyrene, cyclohexane still dissolves the polymer standards at temperatures above 35°C.

Figure 8 shows the calibration curves obtained with TOL and TOL-CHX mixtures. The calibration curve obtained on unmodified silica (column 6) with toluene is also plotted as reference. The shift obtained on the polystyrene bonded silica (column 9) curve when cyclohexane is added to the mobile phase (TOL 80%, CHX 20%) follows the expected trend; however, the fact that the mixture TOL 70%, CHX 30% produces a change in the calibration curve remounting that of pure silica, has to be explained by a partition-adsorption mechanism

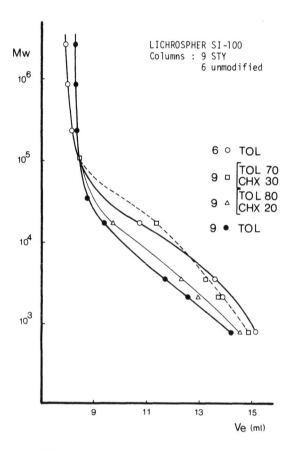

FIGURE 8.   SEC calibration curve shift with toluene-cyclohexane mixtures as the mobile phase.

rather than by a reduction in the swelling of the bonded layer. With
greater than 30% cyclohexane, the polystyrene standards are strongly
delayed or not eluted.

Thus far the achievement of a variable pore diameter packing has
not been successful and seems unlikely to succeed because of the easy
appearance of partition or adsorption effects in GPC when the right
solvent is not used.

ACKNOWLEDGMENTS

This research was carried out during a leave of absence from Dow
Chemical Iberica at the Chemistry Department of the Virginia Polytechnic Institute. I wish to acknowledge the provision of equipment
and facilities by H. M. McNair and the donation of all Lichrospher
packings by M. Gurkin from E. M. Laboratories.

REFERENCES

1. J. C. Moore, J. Polym. Sci., A2, 835 (1964).

2. Waters Assoc. Report DS 044F (1974).

3. Toyo Soda Co., HLC Report 801,1,36 (1974).

4. Showa Denko K. K., Shodex Report.

5. R. Vivilecchia, B. Lightbody, N. Thimot, and H. Quinn, in
    Liquid Chromatography of Polymers and Related Materials I (J.
    Cazes, ed.), Marcel Dekker, New York, 1977, pp. 11-27.

6. K. Unger, R. Kern, M. C. Ninou, and K. F. Krebbs, J. Chromatog.,
    99, 435 (1974).

7. J. J. Kirkland, J. Chromatog., 125, 231 (1976).

8. A. R. Cooper and D. V. Deveer, J. Liquid Chromatog., 1 (5), 693
    (1978).

9. K. Fukano, K. Komiya, H. Sasaki, and T. Hashimoto, J. Chromatog.,
    166, 47 (1978).

10. S. Rokushika, T. Ohkawa, and H. Hatano, J. Chromatog., 176, 456
    (1979).

11. Y. Kato, K. Komiya, H. Sasaki, and T. Hashimoto, J. Chromatog.,
    190, 297 (1980).

12. F. E. Regnier and R. Noel, J. Chromatog. Sci., 14, 316 (1976).

13. H. Engelhardt and D. Mathes, J. Chromatog., 142, 311 (1977).

14. H. Engelhardt and D. Mathes, J. Chromatog., 185, 305 (1979).

15. D. C. Locke, J. Chromatog. Sci., 11, 120 (1973).

16.   P. Roumeliotis and K. Unger, J. Chromatog., 149, 211 (1978).

17.   J. J. Kirkland, J. Chromatog. Sci., 9, 206 (1971).

18.   M. Caude and R. Rosset, J. Chromatog. Sci., 15 405 (1977).

19.   B. Monrabal, in Advances in Chromatography (A. Zlatkis, ed.), 1979, pp. 116-136.

20.   R. H. Casper and G. V. Schulz, in Gel Permeation Chromatography (K. H. Atgelt and L. Segal, eds.), Marcel Dekker, New York, 1971, pp. 225-235.

21.   G. S. Greschner, Makromol. Chem., 180 (11), 2551 (1979).

22.   F. W. Billmeyer, Textbook of Polymer Science, Wiley Interscience, New York, 1971.

23.   G. E. Ham (ed.), Vinyl Polymerization, Volume 1, Part 1, Marcel Dekker, New York, 1967.

24.   A. E. Platt, in Encyclopedia of Polymer Science and Technology (N. M. Bikales, ed.), Volume 13, Wiley Interscience, New York, p. 174.

25.   G. H. Olive and S. Olive, J. Polym. Sci., 48 329 (1960).

26.   C. R. Vogt and W. A. Aue, J. Chromatog. Sci., 16, 268 (1978).

27.   E. N. Fuller, Anal. Chem., 44, 1747 (1972).

28.   J. C. Giddings, Separ. Sci., 5 717 (1970).

29.   P. L. Dubin and K. L. Wright, Polym. Prepr., 15, 673 (1974).

30.   P. L. Dubin, S. Koontz, and K. L. Wright, J. Polym. Sci., B, 15, 2047 (1977).

31.   J. W. Dawkins, Polym. Prepr., 18, 2, 198 (1977).

POLYPHOSPHAZENE POLYMERIZATION STUDIES
USING HIGH PERFORMANCE GPC

G. L. Hagnauer
T. N. Koulouris

Polymer Research Division
Army Materials and Mechanics Research Center
Watertown, Massachusetts

ABSTRACT

The high temperature melt and solution polymerization reactions
of polydichlorophosphazene are investigated. Special techniques are
described for handling and characterizing the hydrolytically unstable
reaction products. High performance GPC and other dilute solution
techniques are used to monitor polymer yield and to analyze polymer
molecular weight, molecular weight distribution, and chain structure.

I. INTRODUCTION

Polydichlorophosphazene [NPCl$_2$]$_x$ is unique as the synthetic precursor

for the preparation of technologically promising poly(organo)phospha-

zenes [1,2]. The polymer (II) may be prepared by either high temper-

ature melt or solution polymerization of hexachlorocyclotriphosphazene

(I) and is reacted with various organic nucleophiles to yield useful

polymers (IV-VI) (see structures below). The rate of polymerization

is quite sensitive to the presence of impurities and increases with

increasing temperature. Melt polymerization reactions are usually

run in sealed, evacuated glass tubes at temperatures between 240 and

255°C and solution polymerizations are best run in solvents with high

dielectric constants ($\varepsilon \geqslant 2$) at about 200°C [3]. Crosslink formation

(III) is unpredictable, tending to increase with increasing

polymerization temperature and polymer yield, and is greatly enhanced by the presence of impurities. Crosslinking is undesirable
since insoluble gel or crosslinked polymer is unsuitable for the
preparation of substituted poly(organo)phosphazenes. To obtain
soluble, high molecular weight polymer (II), the trimer (I) must be
highly pure and polymerization generally must be limited to polymer
yields of 50% or less. The development of polyphosphazenes depends
directly upon optimizing and controlling the polymerization of
polydichlorophosphazene. The polymerization kinetics and the effects of temperature, catalysts, and contaminants on polymer formation need to be investigated. Reaction conditions must be determined
to control polymerization, increase polymer yield, and prevent
crosslinking. Catalyzed polymerizations are desirable to increase
the rate of polymerization and to control polymer structure.

Most attempts to study the $[NPCl_2]_x$ polymerization reaction have
been incomplete. The effects of polymerization conditions on polymer
yield have been investigated, but little attention has been given to
the characterization of polymer structure and to the analysis of
nonpolymeric reaction products. The presence of gel and the fact
that $[NPCl_2]_x$ is moisture sensitive have precluded most attempts to
analyze the polymer's molecular weight (MW) and molecular weight
distribution (MWD). Upon exposure to moisture, the P-Cl bonds

hydrolyze and the polymer gradually degrades to phosphoric acid and
ammonia. In apolar solvents like benzene and toluene, trace amounts
of water result in the formation of P-OH side groups which cause the
polymer molecules to associate through dipolar interactions; and
depending upon the extent of hydrolysis crosslinking may occur
through the formation of P-O-P bonds. In polar solvents like acetone
and dimethyl formamide, the polymer hydrolyzes rapidly, solutions
turn turbid, and a white precipitate is evident within a few hours to
several days.

Recently, techniques have been developed for handling and
characterizing $[NPCl_2]_x$ polymerization products [4,5]. Viscometry,
membrane osmometry, light scattering, and high performance gel permeation chromatography (GPC) were used to characterize MW and MWD and to
analyze low and intermediate MW polymerization reaction products.
Polymers (II) prepared by melt polymerization were found to have high
MWs ($\overline{M}_w \approx 2 \times 10^6$ g/mol) and broad MWDs ($\overline{M}_w/\overline{M}_n \approx 5$). Also, the purity
of the cyclic trimer (I) had a significant effect on polymer yield,
oligomer formation, MW and MWD. In this paper, the techniques are
applied to study the uncatalyzed melt polymerization of purified
trimer (I) and are extended to investigate the catalyzed polymerization in 1,2,4-trichlorobenzene (TCB). The rates of polymerization
are evaluated and the effects of polymerization conditions on polymer
chain structure are discussed. The results of GPC analyses using
different types of high performance columns and different operating
conditions are compared.

II. EXPERIMENTAL

A. Materials and Methods

The trimer (I) was obtained from Ethyl Corp. (Ferndale, MI) and was
purified by vacuum distillation, recrystallization from heptane, and
vacuum sublimation to remove contaminants, hydrolysis products and
higher MW cyclic and linear oligomers. After purification, the
trimer (mp 114°C) was found to be free of impurities to the limits

of detection using differential scanning calorimetry, gas chromatography, and GPC.

ACS reagent grade or Burdick & Jackson (Muskegon, MI) solvents
were used. Precautions were taken during distillation and storage
of the solvents to exclude moisture. The trimer, solvents, and
polymerization reaction mixtures were stored under vacuum and handled
in a dry box under a blanket of dry Argon.

Melt polymerization reactions were run in sealed pyrex ampoules
at 250°C. A specially designed, glass ultrafine (0.8 μm) filter
apparatus was used to filter the molten trimer at about 130°C under
nitrogen pressure into the ampoules. Typically, the ampoules contained 8 g of trimer and were sealed under vacuum (0.005-0.010 mm Hg).
The polymerization was initiated by placing the ampoules in an oven
at 250°C and was terminated at different times by removing the ampoules from the oven. After cooling, each ampoule was opened and
its contents were removed and dissolved in 50 ml toluene. The polymer was precipitated with about 400 ml n-pentane and dried under
vacuum. A Rotavapor-R instrument was used to remove the solvents
from the soluble, low MW fraction. Since sample E contained gel, the
soluble polymer was removed by repeated washings with toluene
followed by treatment with n-pentane to compact the gel and squeeze
soluble components from the crosslinked matrix. After isolation,
the precipitated polymer, low MW fraction, and gel fraction (for
sample E) were weighed to calculate yields.

Solution polymerization reactions were run in a three-neck
round-bottom flask at 210°C with 12 g trimer (I) and 17 ml TCB and
using 0.1 wt. % catalyst. The catalyst was prepared by heating a
9:1 molar mixture of polyphosphoric acid and phosphorus pentoxide at
40°C for 4 hr under a blanket of nitrogen [6]. The polymerization
mixture was stirred continuously and blanketed with nitrogen. Aliquots (ca. 0.3 ml) were removed at intervals of 4, 7, and 10 hr and
diluted with TCB for GPC analysis. Because of the volatility of
TCB, the reaction temperature was lowered to 130°C before removing
each aliquot. The 10, 16, and 24 hr samples were obtained from

separate reactions. Polymer yields and gel content were determined
using precipitation techniques.

B. Dilute Solution Characterization

The polydichlorophosphazene samples were characterized with toluene
as the solvent at 25°C. GPC analyses were also run with TCB and
tetrahydrofuran (THF) as the solvents. The precipitated polymers
were completely soluble and filtered with no difficulty through 0.8
and 5 µm membrane filters. The low MW fractions were dissolved and
analyzed in THF solution. Solutions were prepared and stored in a
dry box except for short intervals during which transfer or injection operations were conducted as required for certain analyses.

Cannon-Ubbelohde dilution viscometers were employed for intrinsic viscosity [η] determinations and number-average molecular weights
$\overline{M}_n$(OS) were obtained using a Mechrolab Model 501 membrane osmometer.
Light scattering measurements were made using a FICA 50 instrument
operated with unpolarized light of wavelength λ_0 = 5461 Å and calibrated with benzene (R_B = 1.58 × 10^{-5} cm^{-1}). The average value of
the refractive index (RI) increment as determined using a Brice-
Phoenix differential refractometer was (dn/dc) = 0.635 ml/g. A
computer program incorporating a polynomial equation for the least-
squares analysis of data and a plotting routine for the construction
of Zimm plots was used to evaluate weight-average molecular weights
$\overline{M}_w$(LS), second virial coefficients A_2 and z-average radii of gyration $<S^2>_z^{1/2}$.

The following conditions were used for GPC analysis:

Apparatus: Waters 150C GPC with 730 Data Module
Samples: $[NPCl_2]_x$ polymers in TCB solution
Concentration: 2µg/µl
Inject volume: 60 µl
Mobile phase: TCB
Temperature: 135°C
Flow rate: 1 ml/min
Columns: DuPont PSM Bimodal 60S + 1000S
Detection: RI, sensitivity 32, attenuation 64
Chart speed: 2 cm/min
Analysis time: 15 min

Apparatus: Waters ALC/GPC-244 instrument with U6K injector,
    R400 RI detector, and Spectra Physics SP4000 data system
Samples: $[NPCl_2]_x$ polymers in toluene or TCB solution
Concentration: 2 µg/µl
Injection volume: 90 µl
Mobile phase: toluene or THF
Temperature: ambient
Flow rate: 1 ml/min
Columns: Waters µBondagel - (2) E-linear + (1) 125Å
Detection: RI 32X, SP4000 plotter attenuation 5
Chart speed: 4 cm/min
Analysis time: 10 min

Apparatus: Waters ALC/GPC-244 instrument with U6K injector,
    R400 RI detector, and Spectra Physics SP4000 data system
Samples: Soluble low MW fractions of $[NPCl_2]_x$ polymerizations
    in THF or in TCB solutions
Concentration: 20 µg/µl
Inject volume: 90 µl
Mobile phase: THF
Temperature: ambient
Flow rate: 2 ml/min
Columns: Waters µ Styragel - (1) 10^3 + (1) 500 + (3) 100Å
Detection: RI 32X, SP4000 plotter attentuation 5
Chart speed: 1 cm/min
Analysis time: 25 min

Discrete area segments were computed over 6 and 4 sec time intervals during polymer elution for the 135°C and ambient temperature
analyses, respectively. The raw data were transferred to a Hewlett
Packard HP9830 computer for evaluation and plotting. Standard methods were applied for integrating the peak areas of the low MW fractions.

II. RESULTS AND DISCUSSION

A. Melt Polymerization

The composition of the polymerization products at different stages
of reaction are given in Table 1. The total wt. % polymer is the
sum of the soluble polymer precipitated with n-petane, the soluble
polymer remaining in the low MW fraction, and the insoluble gel.
The wt. % soluble polymer and other components in the low MW fractions were determined by GPC analysis (Fig. 1). Polymer elutes at
the exclusion limit (12-15 min) followed by high MW oligomers

TABLE 1

Compositional Analysis of Polymerization Mixtures

Sample	Melt Polymerization					Solution Polymerization		
	A	B	C	D	E	F	G	H
Reaction time (hr.)	60	100	150	215	180	10	16	24
Polymer (wt. % sol. ppt.)	10.0	22.6	45.2	47.8	40.5	6.3	21.9	2.9
Polymer (wt. % insol. gel)	0	0	0	0	41.4	0	4.0	48.9
Total polymer (wt. %)	15.1	27.3	49.1	51.3	86.5	16.0	29.9	51.8
High MW oligomers (wt. %)	1.4	0.7	1.2	0.8	2.0	0.3	0.2	--
Cyclic hexamer (wt. %)	3.3	2.2	3.1	3.5	3.0	1.4	1.2	--
Cyclic tetramer (wt. %)	0	0	0	0	0.3	0	0	--
Total yield (wt. %)	19.8	30.2	53.4	55.6	91.8	17.7	31.3	51.8

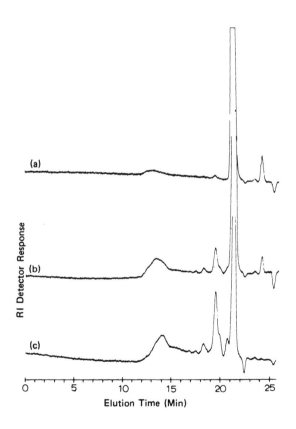

FIGURE 1. GPC analysis of soluble low MW fraction of $[NPCl_2]_x$ melt polymerization (a) Sample A, (b) Sample C and (c) Sample E.

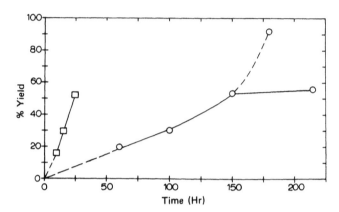

FIGURE 2. Wt% total yield versus polymerization time for melt
polymerization at 250°C (O) and for catalyzed solution polymerization in TCB at 210°C (□).

(15-19 min), cyclic hexamer (19.5 min), cyclic tetramer (20.7 min),
and cyclic trimer (21.5 min). The peak at 24.2 min is due to residual toluene. The yield increases linearly with increasing
polymerization time to 150 hr and then levels off between 150 and
215 hr (Fig. 2). Repeat experiments verify the melt polymerization
behavior between 150 and 215 hr and suggest that sample E is anomalous. The high yield and gel in sample E may be due to the presence
of a contaminant during polymerization. It is also noted that
cyclic hexamer (and perhaps tetramer) and high MW oligomers form
early in the reaction and remain at relatively constant concentration levels to high extents of polymerization. For yields between
20 and 50 wt. %, the reaction rate is about 0.3 wt. %/hr.

The dilute solution parameters in Table 2 were determined by
characterizing the polymer from the precipitated fractions. Sample
E is not representative since nearly half the polymer was insoluble.
Except for sample E, both $\overline{M}_n$ and $\overline{M}_w$ increase with polymerization
time and $\overline{M}_w/\overline{M}_n = 4.0 \pm 0.3$. Classical intrinsic viscosity plots
with normal Huggins constants $0.35 \lesssim k' \lesssim 0.5$ were obtained for
samples A, B, and C. Sample D deviated slightly from ideality and
E behaved anomalously. The large k' value for E is indicative of
polymer association and may be the consequence of P-OH interactions
or perhaps the presents of microgel. The light scattering ratio

TABLE 2

Polydichlorophosphazene Dilute Solution Parameters:
Melt Polymerization

Sample	A	B	C	D	E
Reaction time (hr)	60	100	150	215	180
$[\eta]$ (dl/g)	2.38	2.94	2.99	2.84	1.24
k'	0.40	0.41	0.55	0.77	1.49
$\overline{M}_n$ (OS) $\times 10^{-6}$	0.56	0.56	0.68	0.72	0.57
$\overline{M}_w$ (LS) $\times 10^{-6}$	2.07	2.40	2.55	3.08	1.34
$\overline{M}_z$ (GPC) $\times 10^{-6}$	3.0	4.6	5.0	6.0	6
$\overline{M}_w/\overline{M}_n$	3.7	4.3	3.8	4.3	2.4
$<S^2>_z^{1/2}$ (Å)	830	900	910	820	580
C_∞ (Eq. 4)	26	30	28	27	17
$C_{\infty,z}$ (Eq. 5)	31	23	23	15	8

$<S^2>_z/\overline{M}_w$ is smaller for samples D and E (ca. 0.23) than for A, B, and C (ca. 0.33), suggesting that D and E may contain associated or branched polymer. Finally, second virial coefficients A_2 calculated from the osmometry and light scattering experiments equal zero.

GPC chromatograms for samples A and B are shown in Figure 3. The chromatograms of B, C, and D are quite similar--a bimodal distribution tailing to higher elution times (i.e., a low MW tail). The GPC profile of E was broader and flatter than shown for the polymers in Figure 3. Repeated injections produced identical chromatograms, and similar profiles were obtained for analyses run with toluene as the mobile phase at ambient temperature and with TCB at 135°C. In calibrating the Waters μBondagel and the DuPont PSM Bimodal columns with narrow MWD polystyrene standards, it was noted in both cases that polydichlorophosphazene elutes over a relatively straight region of the calibration plots. Therefore, a two parameter equation,

$$\log_{10} M_i = c_0 + c_1 t_i \tag{1}$$

where c_0 and c_1 are constants, was calculated from the molecular weights M_i and elution times t_i of the polystyrene standards eluting during the same interval as polydichlorophosphazene for both

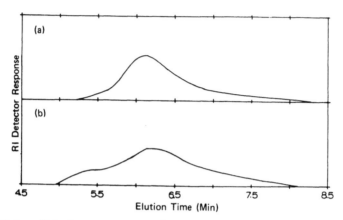

FIGURE 3.   GPC chromatograms of melt polymerized samples (a) A and
(b) B.   Toluene with μ Bondagel columns.

sets of columns; and the polydichlorophosphazene samples were used
as broad MWD standards to transform the equations for polydichlorophosphazene analysis.   The calibration method is described in reference 4.
The calibration equations are

$$\log_{10} M_i = 11.29 - 0.816\ t_i \tag{2}$$

and

$$\log_{10} M_i = 13.35 - 0.929\ t_i \tag{3}$$

for polydichlorophosphazene with toluene (μBondagel) and TCB (Bimodal)
as the mobile phases, respectively, and with the elution time in
minutes.   Plots of the cumulative $C(M_i)$ and differential $F(\log M_i)$
MWDs are illustrated in Figure 4.   Except for sample E, the calculated M_n and M_w values correspond with the absolute MW parameters.
The MWs and MWDs are similar to those reported for derivatized
polyphosphazenes [5].

      Since "a vanishing second coefficient in the virial expansion
of the osmotic pressure" is a requirement for "theta (Θ) point" conditions [7] and since $A_2 \approx 0$ for polydichlorophosphazene in toluene
solution at 25°C, the data in Table 2 may be assumed to be determined at a Θ-point.   Hence the unperturbed dimension of

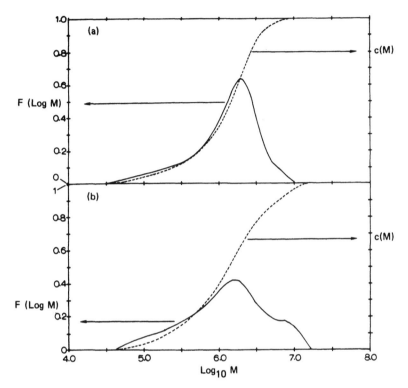

FIGURE 4. Cumulative C(M) and differential F(log M) MWD of melt polymerized samples (a) A and (b) B.

polydichlorophosphazene may be estimated [7] from $[\eta]$ and $\overline{M}_n$ data.

$$C_\infty = \frac{<r^2>_0}{n\ell^2} = \left(\frac{[\eta]_\Theta^{2/3}}{\Phi}\right)\left(\frac{M_b}{\overline{M}_n^{-1/3}\ell^2}\right) \tag{4}$$

or from $<S^2>_z$ and M_z(GPC) data

$$C_{\infty,z} = \frac{<r^2>_{0,z}}{n_z\ell^2} = \frac{6 \cdot M_z \cdot <S^2>_{0,z}}{M_z\ell^2} \tag{5}$$

where $[\eta]_\Theta = [\eta]$ and $<S^2>_{0,z} = <S^2>_z$. C_∞ is the characteristic
ratio, $<r^2>_0$ is the unperturbed mean-square end-to-end distance, n
is the number of skeletal bonds, $M_b = 58$ is the mean MW per P-N
bond, $\ell = 1.6\text{Å}$ is the average P-N bond length, and $\Phi = 0.0026$ dl
$mol^{-1}\text{Å}^{-1}$ is a constant. The subscripts "0" and "z" refer to the
unperturbed state and z-average value.

Calculated characteristic ratios C_∞ and $C_{\infty,z}$ for samples A, B,
and C agree within experimental error suggesting similar chain structure. The low C_∞ value for sample E and the trend for $C_{\infty,z}$ to be
less than C_∞ is indicative of polymer chain branching. The chain
dimensions of a branched polymer are always less than those of a
linear polymer of the same total MW [8]. Considering that insoluble
gel was formed during polymerization, it is likely that sample E
contains soluble branched or crosslinked polymer. Also, since the
probability of branch formation generally increases with increasing
MW, the effect of branching on reducing the value of $C_{\infty,z}$ is expected to be greater than on C_∞. If a polydisperse polymer with
random branching and a large primary chain length is selected as
the branching model, the effect of branching on chain dimensions
may be used to estimate the extent of branching.

$$<g>_z = \frac{<S^2>_{z,br}}{<S^2>_{z,lin}} = \frac{(C_{\infty,z})_{br}}{(C_{\infty,z})_{lin}} \tag{6}$$

where $<g>_z = (1 + Bw/3)^{-1}$ and B_w is the weight-average number of
branched polymer units per polymer molecule. The subscripts "br"
and "lin" denote parameters for branched and linear polymers.
Assuming $(C_{\infty,z})_{lin} \approx 27$, $B_w \approx 2.4$ and 7.1 for samples D and E,
respectively. If a constant branch density is assumed, samples D
and E have 1 branch unit per 11,000 and per 1600 weight-average repeat units, respectively.

The large C_∞ and $C_{\infty,z}$ values for polydichlorophosphazene are
surprising since C_∞ for other polymers is generally in the range
4-10 [7]. C_∞ may be overestimated due to the broad MWD. If branched

polymers are present, calculated parameters M_z(GPC) may be less than their actual values and give rise to inflated values of $C_{\infty,z}$. Finally, $A_2 \simeq 0$ may not be a sufficient condition for assuming a Θ-point due to the broad MWD and possible presence of branched polymer. If Θ-point conditions are not met, the polymer is no longer unperturbed and excluded volume effects become important. Solvent-polymer interactions would perturb chain dimensions and result in overestimated C_∞ and $C_{\infty,z}$ values.

B. Solution Polymerization

The catalyzed solution polymerization proceeds of a rate of about 2 wt. %/hr (Fig. 2). Small amounts of cyclic and higher MW oligomers form early in the reaction, and gel first observed at about 16 hr increases in concentration with polymerization time (Table 1). The GPC chromatogram for the polymer at 10 hr has a similar elution time but is somewhat broader than for the polymer at 4 hr (Fig. 5). The same GPC conditions and calibration (Eq. 3) are used to analyze the solution and melt polymerized polymers. As shown in Figure 6, and Table 3, both $\overline{M}_n$ and $\overline{M}_w$ are greater than 10^6 and the MWDs for the solution polymerized polymers are quite narrow. The similarity of the MW parameters may be a consequence of chain termination when the

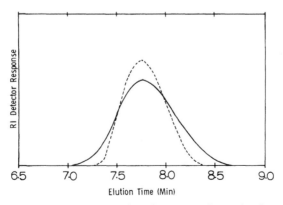

FIGURE 5. GPC chromatograms of solution polymerized polydichloro-phosphazene at 4 hr (------) and at 10 hr (------). TCB with PSM Bimodal columns at 135°C.

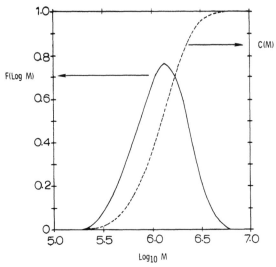

FIGURE 6.   Cumulative C(M) and differential F(log M) MWD of solution
polymerized (10 hr) polydichlorophosphazene.

TABLE 3

Polydichlorophosphazene GPC Analysis - Solution Polymerization

Sample	M_n (10^{-6})	M_w (10^{-6})	M_z (10^{-6})	M_w/M_n
4 hr (TCB)	1.28	1.49	1.71	1.17
7 hr (TCB)	1.33	1.52	1.73	1.15
10 hr (TCB)	1.22	1.78	2.61	1.45
10 hr (THF)	1.27	1.83	2.27	1.43

temperature was lowered to remove aliquots for the 4 hr and 7 hr
analyses.   The broader MWD of the 10 hr aliquot and formation of gel
at longer polymerization times suggest a change in polymerization
mechanism and may be the result of the viscosity increasing with
polymer yield.

Upon evaluating different GPC conditions, it was found that THF
is an acceptable mobile phase with μBondagel columns for analyzing

polydichlorophosphazene dissolved in toluene or TCB solution. THF offers advantages as the GPC mobile phase since it is a better solvent for polydichlorophosphazene and has a larger refractive index increment than TCB or toluene.

In many respects polydichlorophosphazene appears to undergo a "living polymer" polymerization. A high MW polymer with a narrow MWD is formed. The absence of a low MW tail suggests the polymer chains are not being terminated and may be due to the high purity of the cyclic trimer (I). The data also suggests that the rate of initiation must be faster, perhaps because of the nature of the catalyst, than the rate of propagation.

III. COMMENTS

High performance GPC techniques were applied to analyze the reaction products, determine polymer yields, and characterize the MW and MWD of melt and solution polymerized polydichlorophosphazenes. The rate of reaction for the catalyzed solution polymerization at 210°C is significantly greater than the rate of the uncatalyzed melt polymerization at 250°C. High MW polymers are formed in both cases; however, the solution polymerized polymer has a narrow MWD and the melt polymerized polymer has a broad MWD with low MW tail or a bimodal MWD. Cyclic and high MW oligomers are formed early in both reactions and once formed seem to remain at the same concentration level. Dilute solution data suggest that the chain structure of the melt polymerized polydichlorophosphazene samples is uniform and probably linear up to conversions of 50 wt. %. At higer yields branching is indicated and crosslinking may occur especially if impurities are present. The early stage of the catalyzed solution polymerization appears to proceed by a "living polymer" mechanism; however, the MWD broadens when the polymer yield reaches 16 wt. % and an insoluble gel starts to form around 30 wt. %. The broad, bimodal MWD of the melt polymerized polymers suggests a more complex polymerization mechanism.

REFERENCES

1. R. E. Singler, N. S. Schneider and G. L. Hagnauer, Polymer Eng.
    Sci., 15, 321 (1975).

2. R. E. Singler, G. L. Hagnauer and N. S. Schneider, Polymer News,
    5, 9 (1978).

3. G. L. Hagnauer, J. Macromol. Sci.-Chem., A16 (1981).

4. G. L. Hagnauer in Recent Advances in Size Exclusion Chromato-
    graphy, (T. Provder, ed.), ACS Symposium Monograph, 138, Wash-
    ington, D. C., 1980, p. 239.

5. G. L. Hagnauer and R. E. Singler, ACS Org. Coatings & Plastics
    Chem. Div. Preprints, 44, 88 (1979).

6. J. E. Thompson, J. W. Wittman and K. A. Reynard, Horizons, Inc.,
    Cleveland, Ohio, NASA Contract NAS9-14717, April 1979
    (N76-27424).

7. P. J. Flory, Statistical Mechanics of Chain Molecules, John Wiley
    & Sons, New York, 1969, Chapter 2.

8. B. H. Zimm and W. H. Stockmayer, J. Chem. Phys., 17, 1301 (1949).

HIGH PRECISION DETERMINATION OF MOLECULAR WEIGHT
CHANGES IN POLYCARBONATE BY USE OF AN INTERNAL
STANDARD IN ROOM TEMPERATURE GPC MEASUREMENTS

M. Y. Hellman
G. E. Johnson

Bell Laboratories
Murray Hill, New Jersey

ABSTRACT

The use of an internal standard in room temperature GPC
measurements has improved accuracy and precision in obtaining
molecular weight averages. Its effectiveness has been demonstrated
by changing experimental conditions used for analysis of polycarbonate of known molecular weight averages. Variables studied included eluent flow rate and concentration and chemical composition
of the internal standard. One application of this method is reported for the study of hydrolytic degradation of polycarbonate.

I. INTRODUCTION

A study of the hydrolytic degradation of polycarbonate was conducted
by following molecular weight changes using GPC. The conventional
method preestablishes a calibration which can result in greater than
5% variation in molecular weight [1]. This would preclude such a
polycarbonate study, since we wish to follow the reaction at short
time intervals, resulting in molecular weight changes as little as a
few percent. Therefore, a method using an internal standard has
been established in order to obtain the precision required on the
molecular weight measurements.

    The first part of this work describes this method and gives a
detailed analysis of the accuracy and precision in measuring a

standard polycarbonate. Included also are changes in experimental
conditions and use of a variety of internal standards. The second
part reports the application of this method to the analysis of
hydrolytically degraded polycarbonate.

II.   EXPERIMENTAL

A. Materials and Equipment

A polycarbonate of known molecular weight and a hydrolytically degraded polycarbonate were studied. The internal standards tested
were benzene (Fisher ACS grade), benzophenone, phenol, and diphenyl
carbonate all obtained from Aldrich. The solvent used was Waters
UV grade tetrahydrofuran (THF).

All analyses were done on the Waters Model ALC/GPC-244 liquid
chromatograph including 5 microstyragel columns with pore sizes of
10^6, 10^5, 10^4, 10^3 and 500 Å. The elutions were done at room
temperature and eluent was detected with the UV detector. A PDP8
Lab/E (Digital Equipment Corp.) minicomputer was interfaced with
the GPC and was programmed for data acquisition and data reduction.
In addition to the usual molecular weight parameters, a program was
added to obtain the elution volume (V_e) for the peak of an elution
curve.

B.   Internal Standard Calculation

A 0.25% polycarbonate solution in THF which contained a 0.05% concentration of the internal standard was injected and eluted. In a previous publication [2] we have described our method for deriving the
absolute calibration for polycarbonate. This first calibration is
made to fit the known molecular weights. This calibration is

$$\ln \overline{M}_w = C_1 + C_2 V_e \tag{1}$$

C_1 and C_2 are determined from Run 1. The correction for each subsequent run was applied to the $C_2 V_e$ term

$$\ln \overline{M}_w(N) = C_1 + C_2 V_e \left[\frac{\text{Internal Standard Run (1)}}{\text{Internal Standard Run (N)}}\right] \qquad (2)$$

C. Experimental Procedure

The known polycarbonate was run four times on day 1 and three times on day 2 to obtain a measure of precision and accuracy. The effect of flow rate variations was tested by a deliberate change of flow rate to both higher and lower flows. Concentration effects of the internal standard (benzene) were next examined. This was done by changing the concentration by 0.5 and 2.0 times the initial concentration. The choice of internal standard molecular structure was evaluated by investigating internal standards of different chemical compositions. After this thorough testing of the standard procedure, hydrolytically degraded polycarbonate was then analyzed.

III. RESULTS AND DISCUSSION

A most important factor for reliable room temperature GPC molecular weight determinations is a constant flow rate. Flow rate reproducibility must be better than 0.3%, otherwise errors in $\overline{M}_n$ and $\overline{M}_w$ will be approximately ±6% [3]. Many workers assume constant flow rate and measured retention volume; however, even small fluctuations in flow rate may result in large errors when determining average molecular weights of polymers [4]. One source of error in flow rate comes from partially plugged filters. The change may be so subtle that the operator cannot quickly assess the problem. When it is eventually detected, there is a great deal of time lost in repetition of runs.

The internal standard, therefore, serves a two-fold purpose. First, since it corrects the calibration of every run, all data become useful and secondly, the variation of the elution volume for the internal standard can help indicate instrumental problems. This becomes evident in assessing the data in Table 1. The variation of the elution volume for the benzene peak during day 1 runs suggested

			TABLE 1		
		Values of Standard $\overline{M}_n = 1.23 \times 10^4$ g/mole $\overline{M}_w = 3.16 \times 10^4$ g/mole			
		Flow Rate = 2cc/min. Day 1			
Sample No.	Uncorrected $\overline{M}_n \times 10^{-4}$	Corrected $\overline{M}_n \times 10^{-4}$	Uncorrected $\overline{M}_w \times 10^{-4}$	Corrected $\overline{M}_w \times 10^{-4}$	Benzene (V_e)
Standard Run 1	1.227	1.210	3.229	3.175	293.952
Standard Run 2	1.341	1.286	3.426	3.240	293.091
Standard Run 3	1.222	1.159	3.391	3.219	293.176
Standard Run 4	1.287	1.203	3.463	3.232	292.707
	$1.269 \pm .056$	$1.215 \pm .053$	$3.3772 \pm .103$	$3.224 \pm .039$	293.23
			Day 2		
Standard Run 5	1.277	1.275	3.178	3.167	292.877
Standard Run 6	1.236	1.239	3.120	3.129	292.969
Standard Run 7	1.199	1.209	3.102	3.123	293.048
	$1.237 \pm .039$	$1.241 \pm .030$	$3.133 \pm .040$	$3.140 \pm .024$	292.972

an instrument problem. A partially plugged filter was found and
changed. Day 2 was significantly better but the internal standard
still increased precision.

The molecular weight averages in Table 1, day 1, show that the
precision for both weight and number average is improved by using
the internal standard correction. The accuracy for the number
average increased from 3% to 1%, and the weight average from 7% to
2%. On day 2, the uncorrected and corrected data were excellent,
but there was still increased precision and accuracy using the
standard.

Table 2 presents the data resulting from a deliberate change
of flow rate, in both directions. This was done by altering the
pump setting from 2.0 ml/min to 1.9 and 2.1 ml/min. Although this
was only a 5% change in the flow setting, the molecular weights
calculated from the established calibration are off by more than a
factor of 2. However, by applying the internal standard correction,
more accurate values are obtained and the elution runs are made
reliable.

The effect of internal standard concentration was then studied.
The results are summarized in Table 3. The concentration was not
critical as long as it did not interfere with the elution of the
polymer and was within the data collection limits of the computer.

Some researchers have suggested that even further reduction in
molecular weight variations should be achievable if the chemical
composition of the sample and the internal standard were not widely
different [1]. We have tested this by using internal standards of
widely varying structure but still having strong UV absorption at
254 nm. Benzene, phenol, benzophenone and diphenyl carbonate were
selected. Table 4 shows the structures, molecular weights and elution volumes of the standards. Table 5 presents the uncorrected
and corrected molecular weight data. For polycarbonate, all of the
internal standards tested gave satisfactory results. There was no
apparent interaction with the solvent or column packing and all
eluted well after the polymer peak. A typical chromatogram is presented in Figure 1.

TABLE 2				
Values of Polycarbonate Standard $\overline{M}_n = 1.23 \times 10^4$ g/mole $\overline{M}_w = 3.16 \times 10^4$ g/mole				
Flow Rate 2.1cc/min.				
Sample	Uncorrected $\overline{M}_n \times 10^{-4}$	Corrected $\overline{M}_n \times 10^{-4}$	Uncorrected $\overline{M}_w \times 10^{-4}$	Corrected $\overline{M}_w \times 10^{-4}$	Benzene Peak
Standard Run 1	2.759	1.376	6.548	3.217	278.872
Standard Run 2	2.525	1.263	6.355	3.165	279.113
Flow Rate = 2.0cc/min.				
Standard Run 3	1.277	1.275	3.178	3.167	292.877
Standard Run 4	1.236	1.239	3.120	3.129	292.969
Flow Rate 1.9cc/min.				
Standard Run 5	0.755	1.401	1.757	3.169	305.663
Standard Run 6	0.681	1.288	1.750	3.164	305.662

TABLE 3			
2.0 (Benzene Concentration) — Flow Rate 2cc/min.			
Sample	Corrected $\overline{M}_n \times 10^{-4}$	Corrected $\overline{M}_w \times 10^{-4}$	Benzene PK
Standard Run 1	1.298	3.159	293.771
Standard Run 2	1.250	3.150	293.167
0.5 (Benzene Concentration) — Flow Rate 2cc/min.			
Standard Run 3	1.268	3.143	293.169
Standard Run 4	1.183	3.116	293.939

TABLE 4			
Internal Standard	Structure	$\overline{M}_w$	V_e (at 2cc/min.)
Diphenyl Carbonate	$(C_6H_5O)_2$ C=O	214	276
Benzophenone	$(C_6H_5)_2$ C=O	182	282
Phenol	C_6H_5OH	94	279
Benzene	C_6H_6	78	293

HELLMAN AND JOHNSON

TABLE 5					
	Uncorrected $\overline{M}_n \times 10^{-4}$	Corrected $\overline{M}_n \times 10^{-4}$	Uncorrected $\overline{M}_w \times 10^{-4}$	Corrected $\overline{M}_w \times 10^{-4}$	V_e
Standard + Diphenyl Carbonate (1)	1.257	1.27	3.126	3.158	276.07
" (2)	1.256	1.217	3.256	3.157	275.278
Standard + Benzophenone (1)	1.285	1.240	3.224	3.129	282.218
" (2)	1.266	1.21	3.244	3.101	282.059
Standard + Phenol (1)	1.207	1.207	3.163	3.167	278.889
" (2)	1.207	1.272	2.988	3.144	279.863
Standard + Benzene (1)	1.277	1.275	3.178	3.167	292.877
(2)	1.236	1.239	3.120	3.129	292.969

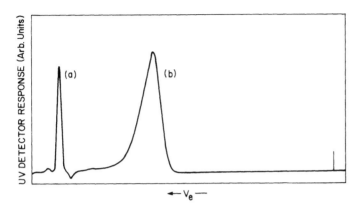

FIGURE 1.   Typical GPC elution curve for polycarbonate in THF using benzene internal standard (a) benzene (b) polycarbonate.

Table 6 presents GPC data for one of the polycarbonates studied. This polymer was hydrolytically degraded at 85°C and 95% relative humidity over a span of 7 months. Samples were taken out at intervals and analyzed by GPC. The analysis also spanned this time. Comparisons required the precise analytical technique of the internal standard. Each time the GPC was used it had to be recalibrated. One run of a known polycarbonate was sufficient to reestablish the starting calibration. The analysis of the degraded sample was then straightforward. An examination of the data from Table 6 shows a variation in the elution volume for the internal standard. However, by using the simple correction method we obtained consistent data showing hydrolytic degradation.

An earlier study [5] on the hydrolytic degradation of polycarbonate utilized the Perkin Elmer GPC (Fig. 2). When using the preestablished calibration, all of the data fell with ±10%. This would have precluded interpretation of hydrolytic degradation. However after correcting the weight averages by our method, the data became meaningful.

TABLE 6

Hydrolytically Degraded Polycarbonate
Lexan 141 85°C 96% RH

Sample	$\overline{M}_n \times 10^{-4}$	$\overline{M}_w \times 10^{-4}$	Benzene PK
Control	1.081	2.624	294.844
306 hrs.	1.048	2.511	292.590
515 hrs.	0.960	2.525	293.949
1019 hrs.	0.947	2.481	293.514
1630 hrs.	1.038	2.470	292.861
2019 hrs.	0.966	2.458	292.795
3140 hrs.	0.9399	2.265	292.817
4220 hrs.	0.856	2.270	291.801
5156 hrs.	0.871	2.152	294.015

In Figure 2, we plot $\overline{M}_w$ vs. log aging time (days) for samples degraded at 95% relative humidity at 50, 60, 71, 85 and 125°C. Included in this figure is a plot of all data shifted to 71°C using a time shift of one decade for every 30°C. Both the original data at 71°C and the data shifted from all temperatures fell on the same curve. This substantiates the effectiveness of our method.

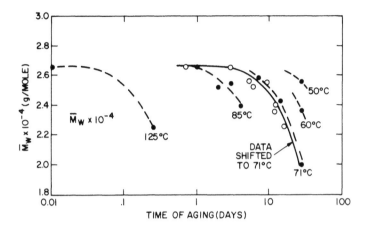

FIGURE 2.   Plot of $\overline{M}_w$ for polycarbonate versus log aging time: ●
original data ○ data shifted to 71°C using Arrhenius parameters
determined from time to obtain an identical degraded $\overline{M}_w$ at various
temperatures.

IV.   CONCLUSIONS

1.   The addition of an internal standard in the GPC analysis of
polycarbonate reduces the effect of flow rate variations. This
improves the accuracy of the weight average molecular weight
measurement from ±7% to ±2% and the number average molecular
weight from ±5% to ±2%.

2.   This method eliminates lengthy recalibrations.

3.   It is time saving in identifying subtle instrument problems
which cause flow rate variations.

4.   Concentrations and chemical composition of the internal standard
are not critical. All of the internal standards studied gave
excellent results.

ACKNOWLEDGMENTS

The authors wish to thank P. C. Kelleher for supplying the degraded
polycarbonate samples and G. N. Taylor for helpful suggestions.

REFERENCES

1. C. N. Patel, J. Appl. Pol. Sci., 18, 3537-3542 (1974).

2. M. Y. Hellman, "Liquid Chromatography of Polymers and Related
    Materials", Vol. 8, 29 (1976).

3. D. D. Bly, H. J. Stoklosa, J. J. Kirkland, and W. W. Yau, Anal.
    Chem., 47, 1810, 2328 (1975).

4. J. V. Dawkins and G. Yeadon, Polymer, 20, 981 (1979).

5. M. Y. Hellman, unpublished results.

A COILED MICROCOLUMN FOR FAST GEL
PERMEATION CHROMATOGRAPHY

C. D. Chow
M. W. Long, Jr.

Analytical Laboratories
Dow Chemical U.S.A.
Midland, Michigan

ABSTRACT

A coiled microcolumn with an inside diameter of 0.04 in. and a
coil radius of 6 in. can be used for fast gel permeation chromatography (GPC). Molecular weight distributions of polymers have been
determined within 30 min with this column. The resolution of the
column packed with controlled-pore glass was comparable to that of
a Waters Associates Model 200 GPC with Styragel column. Molecular
weights of styrene-related polymers obtained both from the fast GPC
with the coiled microcolumn and Waters Associates Model 200 GPC with
straight Styragel columns were in excellent agreement. No deterioration in performance of the coiled column has been observed after more
than 1000 sample runs over a period of 2 years.

I.   INTRODUCTION

Fast gel permeation chromatography (GPC) with an analysis time of

less than 30 min has been the subject of many recent publications

[1-4]. Straight columns with diameters of 2 mm I.D. and up were

used in those studies.

It is common practice in GPC to use straight columns. The ef-

fect on the efficiency and resolution of GPC of coiling the column

was considered detrimental. However, recent studies [5-7] concluded

that there was no loss in efficiency when a small column diameter

(2.4 mm I.D.) and a large coil radius (5.1 cm) or a coil in a figure eight or "S" configuration were used.

The Analytical Laboratories of the Dow Chemical Company has been working with coiled Teflon* and stainless steel microcolumns (less than 2 mm I.D.) since 1965 [8]. A coiled microcolumn has the advantages of easy storage and arrangement, and it saves in the cost of the solvent, tubing and packing materials. The use of a small amount of flammable solvent greatly reduces the fire hazard.

This paper is concerned with the development of the coiled microcolumn for fast GPC while taking advantage of recent innovations in liquid chromatographic theory and techniques. The coiled micro-column was used to determine the molecular weights (MW) of styrene-related polymers. Its resolution and the MW results were compared with those obtained from Styragel[+] columns using a conventional Waters Associates Model 200 GPC.

II. EXPERIMENTAL

A Waters Associates ALC 202/R401 with a volume indicator, glass siphon, and a M-6000 solvent delivery system was used for the study. The glass siphon had a volume of 0.17 ml. A Valco Model CV-6UHPA-C-20 valve, with a 50 µl sample loop, was installed as an injection valve. All connections were made with 1/16 in. O.D. × 0.005 in. I.D. seamless stainless steel tubing, and drilled-through stainless steel unions. 1/16 in.× 1/16 in. stainless steel filter elements with a 5 µm poros-ity were used for column connections to prevent the packing material from escaping or plugging the connecting tubing.

The column used was made of tempered soft-annealed seamless stainless steel tubing, type 304, with the dimensions of 1/16 in. O.D. × 0.04 in. I.D. (1 mm). An equal volume mixture of Electro-Nucleonics controlled-pore glass CPG-HS 2000, 1400, 700, 350, 170

---

*Trademark of DuPont Company
[+]Trademark of Waters Associates.

and 75 Å was used as column packing material. Two sections of 12
ft. columns were precoiled to a radius of 6 ins. The columns were
packed with an aqueous-slurry of CPG-HS as fast as possible using a
Waters Associates M-6000 pump while the columns were immersed in an
ultrasonic bath. The two columns were then combined to make a total
column length of 24 feet.

Polystyrene standards of a narrow molecular weight distribution
(MWD) from Pressure Chemical Co. were used to calibrate the column
at two flow rates: 0.1 and 0.2 ml/min. A concentration of 0.1
weight percent of each narrow molecular weight standard, dissolved
in tetrahydrofuran, was used. The chart speed of a Sargent-Welch
Model DSRG dual pen recorder was set at 0.5 and 0.2 in./min for flow
rates of 0.2 and 0.1 ml/min, respectively. All the molecular weights
reported in this paper are based on polystyrene and were determined
at the flow rate of 0.2 ml/min.

Seven 4 ft. × 3/8 in. Styragel columns were used in the Waters
Associates Model 200 GPC with the following designations: 10^6, 10^5,
10^4, 10^4, 10^4, 10^3, and 10^2 Å. The flow rate was 1 ml/min, and
tetrahydrofuran was used as solvent and eluent for all the work.

III.   RESULTS AND DISCUSSION

The coiled microcolumn was calibrated with mixed polystyrene (PS)
standard solutions. The chromatograms of the standard solutions are
shown in Figure 1, and the calibration curve derived from them is
shown in Figure 2.

The calibration curve of this column was linear over a broad
polystyrene molecular weight range of 3000 to 1,500,000. The calibration curve at both the higher and the lower molecular weight ends
deviated drastically from linearity beyond this range. The molecular
weights were calculated using the linear portion of the calibration
curve.

The resolution values for various polystyrene standard pairs
were calculated using the following equation:

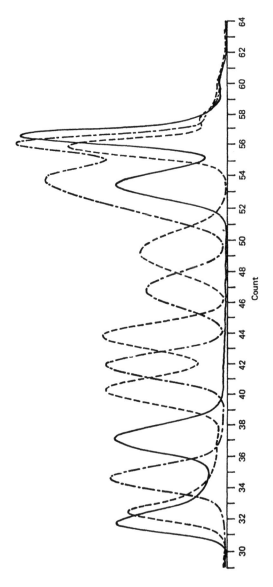

FIGURE 1. GPC curves of polystyrene standards of various molecular weight.

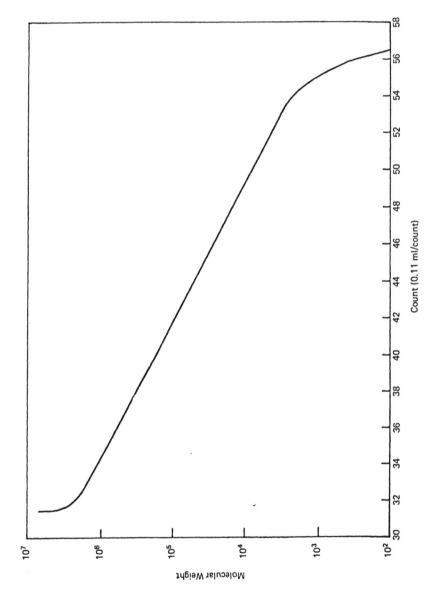

FIGURE 2.   Calibration curve.

$$R = \frac{2(R_2 - R_1)}{W_1 + W_2}$$

where:

R_1, R_2 = the elution volume for peaks 1 and 2, respectively

W_1, W_2 = the base width of peaks 1 and 2, respectively, in the unit of elution volume

The resolution values for the microcolumn are compared with those values obtained from a Waters Associates Model 200 GPC and are shown in Table 1.

It is obvious from Table 1 that the resolution of the fast GPC, using the coiled microcolumn, is equal at the low MW end but is far better at the high MW end than that of the Model 200. The plate counts per foot were 330 and 290 for flow rates of 0.1 and 0.2 ml/min, respectively, and were obtained from an injection of 0.05% trichlorobenzene.

The repeatability of the instrument at two flow rates was excellent, as shown in Table 2. The elution volume of the narrow fraction polystyrene standards became less when the flow rate decreased, especially for the high MW compounds. This is in general agreement with the observation of Little [1].

The MW of a broad molecular weight distribution (MWD) sample PS 1683 was obtained both with the fast GPC and the Model 200 GPC. PS 1683 has a weight average molecular weight ($\overline{M}w$) of 250,000 and a

TABLE 1

Comparison of Resolution

GPC	Column	Total elution time (min)	Flow rate (ml/min)	Pressure (psig)	Resolution of Polystyrene Standard Resins 1,800,000/ 160,000	160,000/ 51,000	51,000/ 10,000
Waters Assocs. Model 200	28' × 3/8" Styragel	350	1	150	1.86	1.24	1.39
Fast GPC	24' × 0.04" CPG–HS	30	0.2	2200	2.46	1.00	1.26

TABLE 2

Reproducibility of Elution Volume at Two Flow Rates

Molecular weight of polystyrene standards	Flow Rate (ml/min)					
	0.2			0.1		
	Run 1	Run 2	Run 3	Run 1	Run 2	Run 3
7,100,000			31.50	31.17		
3,700,000			31.56	31.33		
2,610,000		31.88			31.66	31.74
1,800,000	32.62	32.48	32.52	32.40	32.46	
867,000	34.70	34.65	34.72	34.68	34.68	
411,000	37.21	37.15	37.21		37.10	37.13
160,000	40.35	40.28	40.35	40.21	40.31	
97,200	41.95	41.90	41.91	41.87	41.84	
51,000	43.83	43.85	43.83	43.68	43.79	
19,800	46.85	46.80	46.85	46.82	46.74	
10,300	49.26	49.30	49.35	49.13	49.18	
4,000			52.32	52.20		
3,500	53.42	53.50	53.58		53.36	53.38
2,500	53.77	53.82	53.85	53.75	53.74	
1,140 (squalane)			55.0	54.75		
Trimer	55.92	55.95	56.10	55.95	55.95	
Dimer	56.1	56.1	56.1	56.14	56.07	
Monomer	56.5	56.5	56.77		56.56	56.63

number average molecular weight ($\overline{Mn}$) of 100,000. The $\overline{Mw}$ and $\overline{Mn}$ of PS 1683 were determined by using light scattering spectrophotometry and osmometry, respectively. The GPC results of PS 1683 are shown on the first line of Table 3. The differences between the two GPC results and between the GPC and the light scattering method were within the generally acceptable 5-10% variation of the various techniques.

In some instances, the coiled microcolumn provided a narrower MWD than that obtained from the Model 200. This phenomenon also occurred with dimethyldichlorosilane-treated CPG as the column packing. The authors have suspected that this is due to the adsorption of the ionic polymerized narrow fraction polystyrene standards. Work has been undertaken to study this effect. However, thermally polymerized PS 1683 can be used as the calibration standard [9-10] to solve this problem. The MW and MWD of several hundred polystyrene-related polymers were determined and compared with the results obtained from

TABLE 3

$\overline{M}w$ and $\overline{M}n$ Comparison

Polymer type	$\overline{M}w$		$\overline{M}n$	
	Waters Model 200	Fast GPC	Waters Model 200	Fast GPC
PS 1683	254,000	258,000	100,000	102,000
Polystyrene				
1	195,000	197,000	88,200	92,200
2	219,000	227,000	105,000	119,000
3	237,000	241,000	104,000	99,900
4	252,000	256,000	124,000	134,000
5	283,000	284,000	75,400	71,600
6	293,000	301,000	135,000	139,000
7	305,000	325,000	153,000	171,000
8	326,000	340,000	152,000	172,000
9	348,000	367,000	181,000	202,000
Rubber Modified Polystyrene				
1	192,000	189,000	66,900	73,100
2	204,000	202,000	74,000	76,900
3	217,000	219,000	81,000	83,700
4	228,000	224,000	90,000	91,400
5	230,000	233,000	81,000	91,500
6	259,000	266,000	101,000	104,000
7	312,000	324,000	98,500	86,300
Styrene-Acrylonitrile Copolymer				
1	147,000	140,000	66,800	64,800
2	175,000	174,000	83,000	86,700
3	189,000	186,000	84,400	90,000
Acylonitrile-Butadiene-Styrene Copolymer				
1	145,000	142,000	59,000	52,000
2	170,000	168,000	70,000	66,000
3	220,000	207,000	79,000	73,000

the Model 200 using the broad MWD PS 1683 as the calibration standard. Some of the representative data are shown in Table 3.

The MW of polystyrene and rubber modified polystyrene were in good agreement between the fast GPC and the Model 200 GPC. Slight deviations did occur where the $\overline{M}w$ exceeded 300,000. This deviation was obviously due to the fact that the molecular weight above 300,000 was beyond the linear portion of the calibration curve for the Model 200 GPC but still within the linear portion of the coiled microcolumn.

The MWs of styrene/acrylonitrile copolymers, obtained by fast GPC using regular CPG-HS column packing material, did indicate

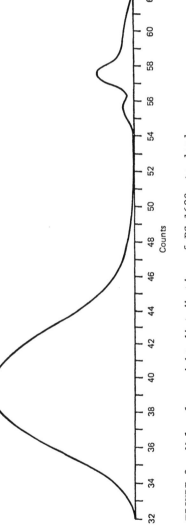

FIGURE 3.   Molecular weight distribution of PS 1683 standard.

inconsistent results, although good agreement was obtained with some samples (Table 3). This inconsistency was presumably due to the adsorption of the very polar cyano group upon the column packing material. Further experimentation will concentrate on the reproducibility of the MW results of the cyano-containing polymers.

More than 1000 samples have been run on the coiled microcolumn, which has been in service for over 2 years. No deterioration in performance has been observed.

ACKNOWLEDGMENT

The authors are grateful to Dr. L. H. Tung and Mr. J. Runyon of the Central Research Laboratory of the Dow Chemical Company for their valuable discussion.

REFERENCES

1. J. N. Little, J. Polymer Sci., A-2, 7, 1775-1783 (1969).

2. B. J. Gudzinowicz and K. Alden, J. Chromatog. Sci., 9, 65-71 (1971).

3. E. P. O'Tocka, J. Chromatog. Sci., 76, 149-157 (1973).

4. Y. Kato, S. Kido, M. Yamamoto, and T. Hashimoto, J. Polymer Sci., 12, 1339-1345 (1974).

5. H. Barth, E. Dallmeier, and B. L. Karger, Anal. Chem., 44, 11, 1726 (1972).

6. L. R. Whitlock, R. S. Porter, and J. F. Johnson, J. Chromatog. Sci., 10(7), 437 (1972).

7. W. Heitz, J. Chromatog., 83, 223-231 (1973).

8. E. G. Owens, J. G. Cobler, J. R. Runyon, and D. E. Zahm, (private communications).

9. B. R. Loy, J. Polymer Sci., 14, 2321 (1976).

10. T. D. Swartz, D. D. Bly, and A. S. Edwards, J. Appl. Polymer Sci., 16, 3353 (1972).

CHARACTERIZATION OF OLIGOMERS BY GPC

W. Heitz

Fachbereich Physikalische Chemie
Polymere
Philipps-University
D-3550 Marburg
Federal Republic of Germany

SUMMARY

Using soft gels and a coiled column system, a high resolution in the separation of oligomers can be obtained. Defects at the endgroups or within the oligomeric chain result in GPC curves typically different. The probability of chain defects increase with chain length. This is shown with epoxy resins. Quantitative information about oligomers needs a correction of detector sensitivity. An equation by Vogl $n = \Sigma \, R_v/M$ with R_v = tabulated group contributions allows us to calculate the refractive index.

The telomerization of THF by Ac_2O using $HSbF_6$ as an initiator results in a PTHF with two ester endgroups. The reaction is characterized by a fast monomer consumption to equilibrium and a slow equilibration under total consumption of Ac_2O. The equilibrium constants of each addition step can be obtained by GPC. Cycles formed in this reaction have been isolated. Cyclic PTHF can be well separated from the linear polymer with OH endgroups.

The synthesis of telechelics by radical polymerization not only is limited by disproportionation reaction, but also by reaction of primary radicals with the oligomer. With aromatic polymers the resulting reaction is not a transfer reaction, but a radical aromatic substitution as shown with styrene and azo(bis-i-butyrate).

I. INTRODUCTION

GPC is a powerful tool to obtain information about irregularities in an oligomeric chain.

To obtain this information we must use the ultimate resolution possible in GPC. The resolution is governed by well-known facts. Independent of the types of equation (Van Deemter [1] or Giddings [2]) describing the separation for a given set of columns, the reduced separation efficiency is described by a master curve (Fig. 1) [3]. The plate count is improved by low flow rates and particle diameters. Under given experimental conditions the reduced velocity varies due to the dependence of the diffusion coefficient from molecular weight. In the most important region of oligomer separation-- from monomer to about decamer--it causes a change in diffusion coefficient by a factor of 3 to 5. A separation of oligomers is normally expanded over the range of abscissa values of Figure 1. There is no risk that we choose the rate of elution so low that free diffusion is an important factor for worsening of the separation, e.g., that we obtain reduced velocities well below the minimum.

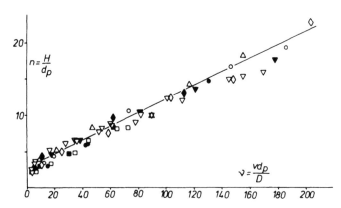

FIGURE 1. Reduced separation efficiency of a poly(vinyl acetate) gel (crosslinked with 5 mol % butanediol divinyl ether); d_p = 0.0191 cm - open symbols; d_p = 0.0382 cm - full symbols; □ ■ Benzene; ▽ ▼ m-Bitolyl; △ ▲ Tetramethyl-p-quaterphenyl; ◇ ♦ Tetramethyl-p-quinquephenyl; ○ ● Octamethyl-p-octiphenyl.

II.   SEPARATION CONDITIONS

In GPC the separation is limited to a volume which is a fraction of
the column.   The actual figure depends on the gel to be used.   That
is why we still use gels of low crosslinking density.   These gels
are prepared by copolymerizing a monomer with a small amount (0.5-
2%) of a crosslinking agent.   These gels have no porosity in the
dry state.   They form a permeable, loose network on swelling,   The
amount of none permeable matrix necessary to fill the space is small
and thus up to 0.75 v_T (v_T = total volume of the column) can be used
for the separation.   This gain in volume can merely be made up by an
increase of plate count as obtained by a decrease in particle size.
In fact the mechanical properties of these gels are rather poor.
There is an upper pressure in the application of these gels during
packing and use of a column.   Exceeding this limit (usually 10-20
bars) results in an irreversible breakdown of the plate count.
   The use of a coiled column system [4] allows an easy handling
of long column systems.   The columns are made from poly(tetrafluoro
ethylene) tubing.   Steel tubing is not as effective with respect to
plate counts obtainable.   The best values we obtained by this method
was a plate count of 160,000 [Conditions: column 0.2 × 1000 cm; DMF,
poly(vinyl-acetate) gel; d_p = 19 μm (dry), 27 μm (swollen)].   During
8 years continuous·use the plate count of this column decreased to
90,000.

III.   QUALITATIVE INFORMATION

GPC can give information about two kinds of irregularities in an
oligomeric chain:

1.   Differences in the structure of the endgroups.

2.   Defects along the chain.

These two chain irregularities cause chromatograms of typically
different appearance.   If the product contains several polymerhomo-
logous series with the same main chain but differing in the

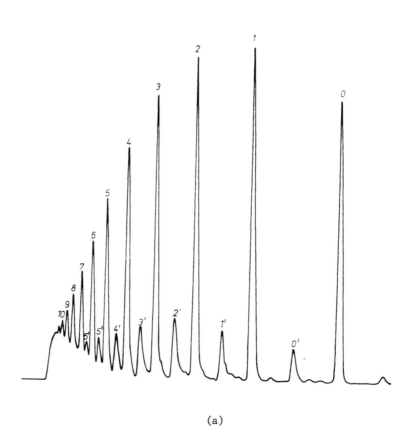

(a)

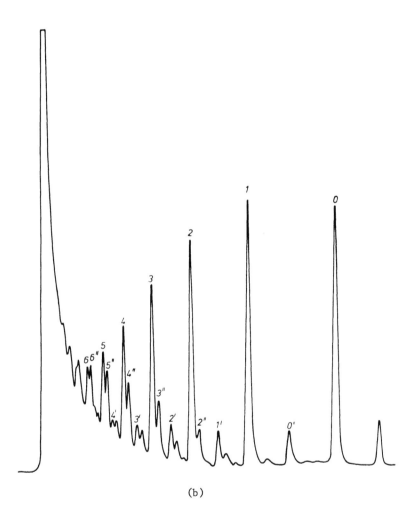

(b)

FIGURE 2 (left and above).   Gel chromatogram of epoxy resins.

The figures at the peak correspond to n in the given formula.   (a)
Prepared at 75°C; (b) prepared at 90°C.

endgroups, then the intensity ratio of the different peak series is almost constant over the range separable. Side reactions may cause irregularities at the monomer units along the chain. The probability of such a defect will increase with chain length. This causes the minor peak series to become the major peak series with increasing molecular weight. This is illustrated with epoxy resins as an example (Fig. 2). If the resin is prepared at a lower temperature (75°C), about 85% of the product has the correct structure (Fig. 2a). The peak series of the second highest intensity is caused by a defect at the endgroups. Braun et al. [4] assigned this structure to the polymer homologous series with one epoxy and one epichlorohydrin endgroup. The molecular weight distribution of both series, e.g., the peak ratio of corresponding peaks is about the same. If the epoxy resin is prepared according to the literature [5], at 90°C side reactions are unavoidable (Fig. 2b). The peak series as given above are present in the same intensity ratio. At this higher reaction temperature the OH-groups within the chain are involved in a reaction. We assume this to be the reaction

$$\sim\!\!\sim\!\!-O-CH_2-CH-CH_2-O\sim\!\!\sim + \;CH_2-CH-CH_2-O\;\sim\!\!\sim$$
$$\qquad\qquad\qquad |\qquad\qquad\qquad\qquad \backslash\!\!\diagup$$
$$\qquad\qquad\qquad OH\qquad\qquad\qquad\qquad\; O$$

$$\downarrow$$

$$\sim\!\!\sim\!\!-O-CH_2-CH-CH_2-O\sim\!\!\sim$$
$$\qquad\qquad\qquad |$$
$$\qquad\qquad\qquad O$$
$$\qquad\qquad\qquad |$$
$$\qquad\qquad CH_2-CH-CH_2-O\sim\!\!\sim$$
$$\qquad\qquad\qquad\quad |$$
$$\qquad\qquad\qquad\quad OH$$

In agreement with this explanation is the fact that n and n" (with the same number of bisphenol A units) have similar elution volumes. The branched molecule being more compact is eluted later. The series n' is also accompanied by satellites the intensity of which increases with molecular weight. Branched molecules with 6 bisphenol A units are present in equal amounts compared to linear molecules of the same size.

IV.   QUANTITATIVE INFORMATIONS

Differential refractometers are the most often used detector systems.
At higher molecular weights the dependence of sensitivity from molecular weight can be ignored.   In the oligomeric field it mainly depends
on differences of refractive indices between solute and solvent if a
correction is necessary.

Refraction indices of some oligomers are tabulated [6].   They
can be calculated by the Lorentz/Lorenz equation

$$n = \left[(1 + 2\frac{\Sigma R_{LL}}{V})/(1 - \frac{\Sigma R_{LL}}{V})\right]^{1/2} \tag{1}$$

where

n = refractive index

R_{LL} = group contribution in the LL-equation

V = molare Volume

which has its theoretical basis in the electromagnetic wave theory
of light.   As Van Krevelen [7] in a critical survey pointed out the
much more simple equation of Vogl [8]

$$n = \frac{\Sigma R_v}{M} \tag{2}$$

where

R_v = group contribution in the Vogl equation

M = molecular weight,

results in values with the same standard deviation (0.4%).   Plots of
n versus $1/P$ are not linear (Fig. 3a), whereas in agreement with Eq.
2, n versus $1/M$ results in a linear relationship (Fig. 3b).   Taking
into account that $P = (M - M_e)/M_m$, Eq. 2 can be transformed to a
linear relationship (Eq. 3):

$$n = \frac{(R_vm)P + (R_ve)}{M} = \frac{(R_vm)}{M_m} - \left[\frac{(R_vm)M_e}{M_m} - (R_ve)\right]\frac{1}{M} \tag{3}$$

where

$(R_v m)$, $(R_v e)$ = group contribution of monomer unit and endgroup
                resp.

M_m, M_e = molecular weight of monomer unit and endgroup resp.
Curves calculated for dinitrile telechelic oligobutadiene and poly
(ethylene oxide) by use of values given in Table 1 are in good agreement with experimental values. In some cases group contribution may
be dependent upon the adjacent group. The adjustment of the straight
line to the n values of the polymer and P = 1 (with endgroups) is a
good fit for all the other oligomers.

The molecular weight distribution obtained with oligomers is
most often narrower than expected from Schulz-Flory or Poisson
distribution resp.. Oligomers obtained by equilibration or radical
polymerization resp. may have M_w/M_n of 1.3. This is due to the fact
that the long chain approximation cannot be applied in this molecular
weight range. But special cinetic and/or thermodynamic effects may
be also of importance. In some cases the polymerization degree 1 is
practically not formed. There exists a smallest possible degree of
polymerization P_0. Thus the probability of growth p will be

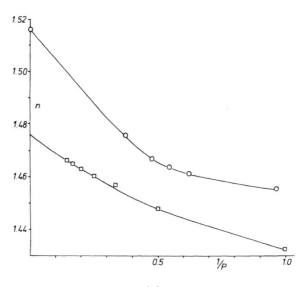

(a)

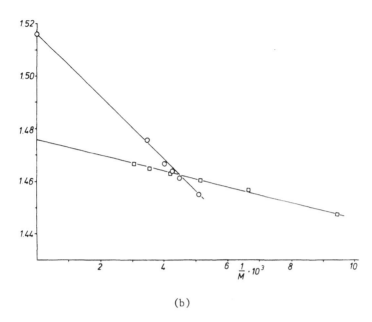

(b)

FIGURE 3 (left and above). Refractive index of oligomers.

$$N \equiv C-\underset{\underset{CH_3}{|}}{\overset{\overset{CH_3}{|}}{C}} \left(CH_2-CH=CH-CH_2 \right)_n \underset{\underset{CH_3}{|}}{\overset{\overset{CH_3}{|}}{C}} - C \equiv N \quad \text{exp. values from [9]}$$

$$HO (CH_2\ CH_2\ O)_n H \qquad\qquad\qquad \text{exp. values from [6]}$$

(a) Refractive index n versus 1/P; (b) n versus 1/M; curves calculated according to Eq. 2 using values of Table 1 giving.

$$n(PB) = \frac{81.9P + 194.18}{54P + 136}; \quad n(PEO) = \frac{63.02P + 23.57}{44.05P + 18.01}.$$

TABLE 1

Group Contributions Used to Calculate the
Refractive Index According to Equation 2

CH_3-	17.66	$-CH-$	23.49	$-OH$	23.57
$-CH_2-$	20.64	$-C\equiv N$	36.67	i-propyliden group	-0.2
$\diagup\!C\!\diagdown$	25.3	$-O-$	23.73	double bond	-6.36

$$P_n - P_0 = \frac{p}{1 - p} \tag{4}$$

M_w/M_n is given by

$$\frac{M_w}{M_n} = 1 + p = \frac{2P_n - 2P_0 + 1}{P_n - P_0 + 1} \tag{5}$$

Telomerizing THF by Ac_2O the polymerization degree 1 is practically not formed. Taking into account this smallest possible degree of polymerization $P_0 = 2$ the calculated molecular weight distribution is in agreement with that found by GPC (Table 2).

TABLE 2

Distribution of $Ac(OCH_2CH_2CH_2CH_2)_n$ OAc and
Equilibrium Constants of the Successive Equilibrium
Constants K_n in Cationic Polymerization of THF
($10°C$; CH_2Cl_2; $HSbF_6$); $[M]_\infty = 2.48$ mol dm^{-3}

P	N_x	Found	$n_x \cdot 10^2$ Calc. $p = 0.5535$[a,b]	$p = 0.6914$[c]	K_n
1	0.0806	0.948	--	30.65	
2	3.5865	42.185	44.65	21.25	17.9
3	2.1660	25.477	24.71	14.74	0.24
4	1.2079	14.208	13.67	10.22	0.23
5	0.6924	8.144	7.57	7.08	0.23
6	0.3295	3.876	4.19	4.92	0.19
7	0.1785	2.100	2.32	3.41	0.22
8	0.1007	1.184	1.28	2.36	0.23
9	0.06160	0.7246	0.71	1.64	0.25
10	0.04393	0.5167	0.39	1.14	0.29
11	0.02776	0.3265	0.22	0.79	0.16
12	0.01675	0.1970	0.12	0.55	0.24
13	0.00956	0.1125	0.07	0.38	0.23
$\Sigma = 8.5017$					

Notes: $P_n = 3.24 \Rightarrow M_n$ (with endgroups) $= 324$ (found $= 337$);
$P_w = 4.15$; $M_w/M_n = 1.28$.

[a] $n_x = p^{x-2}(1 - p)$.

[b] From Eq. 4.

[c] From $P_n = 1/(1 - p)$.

A.   Telomerization of THF

The value of a poly-THF is dependent on its functionality.
Bifunctionality is a basic requirement of useful products.   But the
technical synthesis of products with two OH-endgroups is not without
problems.

In Figure 4 GPC separations of technical poly tetrahydrofurane
samples are compared.   The good base line separation obtained in
Figure 4a together with the analytical informations indicates a correct structure with all the molecules having two OH endgroups.

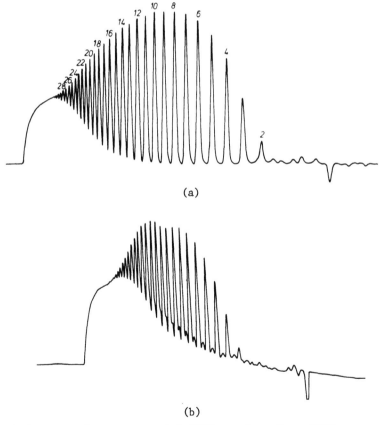

(a)

(b)

FIGURE 4.   GPC of two commercial PTHF samples; M_n = 1000
$HO(CH_2CH_2CH_2CH_2O)_nH$; figures at the peaks indicate n; GPC cond.:
Merckogel 6000, DMF, column: 0.2×1000 cm. flowrate: 0.75 ml/h.

Figure 4b is the characteristic picture of a product containing a
fraction of molecules with endgroup defects.

The OH-groups in question are introduced by an appropriate
initiation and termination step. Initiating the THF polymerization
by a strong protonic acid the first step is a protonation of THF and
by a subsequent nucleophilic attack of a second THF molecule the
first OH-endgroup is formed. The second OH-endgroup is introduced
by reaction of the living chain-end with water or OH^-. Thus, per
mole of oligomer formed one mole of initiator is consumed. By use
of acetic anhydride as transfer agent, only catalytic amounts of a
protonic acid are necessary.

$$\overset{\oplus}{\wedge\wedge\wedge-O} \bigcirc \ + \ Ac_2O \longrightarrow \wedge\wedge\wedge-OCH_2CH_2CH_2CH_2-\overset{\oplus}{O}-Ac_2$$

$$\overset{THF}{\longrightarrow} \wedge\wedge\wedge-OCH_2CH_2CH_2CH_2OAc \ + \ Ac-\overset{\oplus}{O}\bigcirc$$

By this transfer reaction two ester endgroups are introduced. Acetic
anhydride reacts with the oxonium ion forming a diacyloxonium ion. A
subsequent nucleophilic attack of THF results in an ester endgroup
and an acyldialkyloxonium ion, initiating a new chain under formation
of an ester endgroup. The transfer constant can be obtained from low
conversion experiments [10].

The low transfer constant of 0.058 expresses the fact that THF
is a much better nucleophile than Ac_2O. Due to this low transfer
constant THF is consumed in a cinetically controlled reaction much
faster than Ac_2O. But approaching the equilibrium concentration of
THF the net conversion of THF is small, and now the thermodynamically
controlled equilibration dominates (Fig. 5). This results in an
increase of molecular weight with reaction time at low conversions.
The acetic anhydride is consumed in a slow reaction with the consequence of reducing molecular weight. Finally the molecular weight
reaches a limiting value which is the higher, the higher the ratio
of THF/Ac_2O is. This limiting P_n is given by

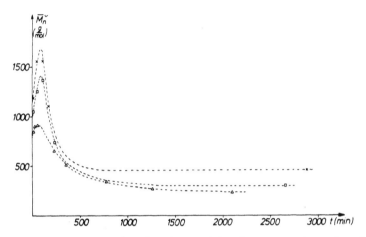

FIGURE 5.   Dependence of molecular weight from reaction time in the
telomerization of THF by Ac_2O [11].
$[THF]_0/[Ac_2O]_0$ = 12.6 x
                  = 6.4 □
                  = 4.3 △

$$P_n = \frac{\Delta[THF]}{[I]_0 + [Ac_2O]_0}$$ (6)

Ac_2O is completely consumed in this reaction.   There is no difference in the product if Ac_2O is added before starting the polymerization or after polymerization of THF.

The first step in the polymerization of THF in presence of Ac_2O
can be a protonation of Ac_2O or THF resp.   The later case can give
rise to molecules with one OH and one ester endgroup.   Figure 6 shows
the GPC traces of products obtained after different reaction times.

Specially at short reaction times a double peak series is observed, one series is caused by a polymer homologous series with one
OH and one AcO endgroup.   Having consumed all the protons of the
initiator, new molecules can only be formed by a transfer reaction
with Ac_2O, i.e., now molecules with two ester endgroups are formed
as shown by the relative increase of the corresponding peak series
with time.   In accordance with the GPC results, the ester functionality as determined by titration increases with reaction time.   The GPC
curves of Figure 6 again show that the molecular weight is reduced in

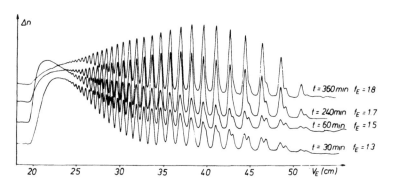

FIGURE 6. GPC of PTHF obtained by initiation with $HSbF_6$ and polymerized in presence of Ac_2O after different reaction times; the ester functionality f_E is obtained by titration.

the equilibration process. Comparing the products after 30 min and 6 h, the high molecular weight fraction has nearly vanished.

Figure 7 shows a GPC of a product of low P_n after reaching the equilibrium conditions. By correcting for the detector sensitivity the mass distribution is obtained. The molecular weight distribution under equilibrium conditions is a Schulz-Flory distribution. Of special interest is that the polymerization degree one is present only in very small amounts. There are chemical reasons for this.

$$\sim\sim\!\!-\underset{\llcorner_____\lrcorner}{O\text{-}CH_2CH_2CH_2CH_2\text{-}O}\overset{\oplus}{\bigcirc} \;\rightleftharpoons\; \sim\!\!\overset{\oplus}{O}\bigcirc \;+\; THF \qquad (7)$$

$$\underset{\llcorner_____\lrcorner}{Ac_2O\text{-}CH_2CH_2CH_2CH_2\text{-}O}\overset{\oplus}{\bigcirc} \;\rightleftharpoons\; Ac\text{-}\overset{\oplus}{O}\bigcirc \;+\; THF \qquad (8)$$

In contrast to the usual propagation/depropagation reaction where an ether oxygen attack has to be considered (Eq. 7), the depropagation of P_2^* involves a nucleophic attack of an ester oxygen (Eq. 8).

The polymerization of THF is an equilibrium polymerization, the equilibrium concentration of which is given by

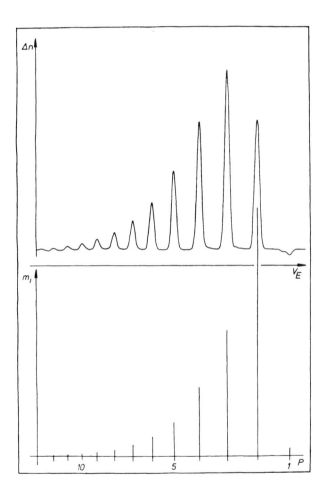

FIGURE 7. GPC of AcO[CH$_2$CH$_2$CH$_2$CH$_2$O]$_n$Ac after equilibration; lower part: mass distribution.

$$K_n = \frac{[P_n^*]}{[P_{n-1}^*][M]_\infty}$$

where

   $[P_n^*]$ = concentration of active species of polymerization degree
        n

   $[M]_\infty$ = equilibrium monomer concentration.

   Usually in relating the equilibrium concentration with the
equilibrium constants the difference in concentration of two neighboring peaks is neglected. At small P_n this is not possible, and so
a dependence of the THF equilibrium concentration from the polymerization degree is obtained. Calculation of the equilibrium constant
requires the knowledge of two neighboring active species. If we
assume that the distribution of active species is the same as in the
final product, this ratio can be measured from the intensity of two
neighboring peaks by GPC. Arguments that this assumption is correct
are given elsewhere [12]. In Table 2 the values of the equilibrium
constant for each addition step are given. The error in the
determination of K_2 is large due to the small difference in refractive index between P_1 and DMF. The average K_n is 0.23. There is no
indication of a dependence of K_n from the degree of polymerization
with the exception of K_2.

   Pruckmayer et al. [13] have shown that cycles may be formed in
the polymerization of THF. In order to obtain knowledge about the
formation of cycles in the telomerization of THF the ester endgroups
were hydrolyzed and the OH endgroups reacted with a diisocyanate.
Linear PTHF is thus converted to a high molecular weight material
with urethane groups. Exhaustive extraction of this polymer with
hexane allows a complete separation of the cycles. Figure 8a shows
a GPC of the fraction of cyclic PTHF which amounts to less than 1% of
a product polymerized at 30°C by bulk polymerization [12]. The poly-
mer-homologous series of cycles are in some molecular weight ranges
well separated from $HO(CH_2CH_2CH_2CH_2O)_n H$ (Fig. 8b).

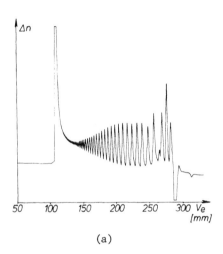

(a)

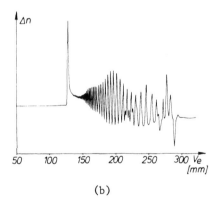

(b)

FIGURE 8. GPC of PTHF; GPC conc. as given in Figure 4. (a) Fraction of cyclic PTHF separated from a telechelic PTHF (see text); (b) artificial mixture of cyclic PTHF and $HO[CH_2CH_2CH_2CH_2O]_nH$.

B. Radical Polymerization of Styrene

A successful synthesis of telechelics by radical polymerization is
mainly dependent from the mode of termination. Only a combination
reaction results in telechelics. Whereas in polymer formation the
termination occurs between two macroradicals, a considerable fraction of molecules are produced by primary radical termination in
the synthesis of telechelics.

The termination in the polymerization of styrene is commonly
considered to be exclusively a combination [14]. Recent results
[15,16] indicate a 20% disproportionation. Primary radical termination may give higher fractions of combination [17]. The reaction of
styrene with AIBN results in dinitrile telechelic oligostyrene. One
of the reasons is the limited solubility of AIBN in aromatic solvent
which limits the investigation range. Using azo-(bis-methyl-i-
butyrate) as an initiator the limits of the statement--functionality
equal to two--are reached. Table 3 gives some results. With increasing molecular weight the tendency to lower functionalities is
obvious. At high initiator to monomer ratios (Table 3 nos. 1, 4, 5)
the functionality exceeds two. Primary radicals ($CH_3OOC-\overset{\displaystyle CH_3}{\underset{\displaystyle CH_3}{C\cdot}}$) attack

TABLE 3

Polymerization of Styrene with AIBM in Toluene at 98°C
(Reaction Time = $10t_{1/2})^a$

No.	$[I]_0$	$[M]_0$ (mol/l)	$[M]$	$[M]_0/[M]$	$M_n{}^b$	P_n	$f_E{}^c$
1	0.1740	0.8620	0.2552	3.377	630	4.16	2.11
2	0.1534	1.777	0.5274	3.369	1080	8.42	1.90
3	0.1245	3.089	0.8117	3.806	2100	18.21	1.86
4	0.3864	1.9873	0.3028	6.562	787	5.42	2.17
5	0.8312	4.1436	0.1223	33.88	888	6.58	2.23

[a] $t_{1/2}$ = 15 min.

[b] By vapour pressure osmometry.

[c] Ester functionality obtained by $f_E = \dfrac{\%O \cdot M_n}{3200}$.

the aromatic ring of polystyrene. By radical substitution a fraction
of molecules with three ester groups are formed. Figure 9 demonstrates that the polymerization of styrene initiated by azo-(bis-
methyl-i-butyrate) is not to describe by a simple reaction scheme.

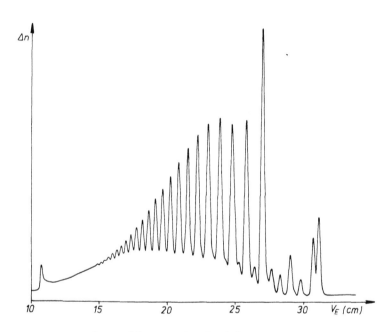

FIGURE 9.   GPC of an oligomer obtained from styrene an

$$CH_3-O-\overset{\overset{O}{\|}}{C}-\overset{\overset{CH_3}{|}}{\underset{\underset{CH_3}{|}}{C}} - N=N-\overset{\overset{CH_3}{|}}{\underset{\underset{CH_3}{|}}{C}} - \overset{\overset{O}{\|}}{C}-O-CH_3$$ (Example 1 from Table 3).

REFERENCES

1. J. J. Van Deemter, F. J. Zuiderweg, and A. Klinkenberg, Chem.
    Eng. Sci., 5, 271 (1956).

2. J. C. Giddings and K. L. Mallik, Anal. Chem., 38, 997 (1966).

3. W. Heitz and J. Coupek, J. Chromatog., 36 290 (1968).

4. D. Braun and D. W. Lee, Angew. Makromolekulare Chem., 51, 11 (1976).

5. D. Braun, H. Cherdron, and W. Kern, Praktikum der makromolekularen organischen Chemie, 3rd ed., Hüthig Verlag, Heidelberg, 1979.

6. M. Rothe, in Polymer Handbook (J. Brandrup and E. H. Immergut, eds.), Wiley, New York, 1975.

7. D. W. Van Krevelen, Properties of Polymers, 2nd ed., Elsevier, Amsterdam, 1976.

8. A. I. Vogel, W. T. Cresswell, G. H. Jefferey, and J. Leicester, Chem. a. Ind., 376, 1951.

9. P. Ball, Thesis, University Mainz, 1977.

10. W. Stix and W. Heitz, Makromolekulare Chem., 180, 1367 (1979).

11. W. Stix, Thesis, University Mainz, 1977.

12. H. J. Kress, W. Stix and W. Heitz, Makromolekulare Chem. (in press).

13. I. M. Robinson and G. Pruckmayr, Macromolecules, 12, 1043 (1949).

14. G. G. Eastmond, in Free-Radical Polymerization, Vol. 14A, from Comprehensive Chemical Kinetics (C. H. Bamford and C. F. H. Tipper, eds.), Elsevier, Amsterdam, 1978.

15. O. F. Olaj, J. B. Breitenbach, and B. Wolf, Monatsh. Chem., 95, 1646 (1964).

16. K. C. Berger and G. Mayerhoff, Makromolekulare Chem., 176, 1983 (1975).

17. K. C. Berger, Makromolekulare Chem., 176, 3575 (1975).

18. G. Gleixner, O. F. Olaj, and J. W. Breitenbach, Makromolekulare Chem., 180, 2581 (1979).

19. W. Konter, B. Bömer, K. H. Köhler, and W. Heitz, Makromolekulare Chem. (in press).

LIQUID CHROMATOGRAPHIC CHARACTERIZATION OF PRINTED CIRCUIT BOARD MATERIALS

Deborah K. Hadad

Lockheed Missiles and Space Company, Inc.
Sunnyvale, California

ABSTRACT

Current purchase specifications for resin impregnated composite materials (prepregs) used in printed circuit board fabrication are based on performance criteria and apply to a variety of commercially available materials. Unfortunately, several materials may meet the performance requirements but process quite differently in actual production due to differences in their chemical formulations. Thus, in order to improve the capability to predict ultimate performance of raw materials and consistently produce high quality hardware, composition specifications which guarantee that each lot of material is chemically equivalent must be instituted. A study utilizing liquid chromatography to solve such a problem is discussed.

I. INTRODUCTION

Although chemical characterization techniques are beginning to be utilized for incoming quality control of printed circuit board (PCB) raw materials [1-2], the large majority of aerospace manufacturers still purchase these materials to military specifications based primarily on performance criteria. Because a substantial amount of effort has been expended in studying and developing chemical quality assurance test methods for structural advanced composite prepregs [3-8], it was only logical to extend these techniques to PCB prepregs.

157

The production processes to produce rigid flex PCBs were
established using material meeting a military specification, MIL-G-
55636. This specification defines material classes according to
their flow characteristics with the minimum flow (Class I) resin
having a flow of 20% or less. Because of Lockheed's processing and
hardware requirements, this specification was modified to include a
material with 2% or less flow. Vendors qualifying under this
specification had to meet this flow requirement as well as all the
performance tests included in the MIL spec.

The object of a purchase specification is to test potential raw
materials and guarantee that those substances can be used successfully in fabrication before committing them in actual hardware.
Unfortunately, under the current purchase specification for printed
board prepreg, it is possible to meet both flow and performance requirements with matrix resins that are chemically different and
which process quite differently. The potential for a collection of
processing problems is obvious if vendors are free to change their
formulations at will. The result of just such a lack of chemically
definitive incoming material control tests is illustrated in the
following discussion.

II. BACKGROUND

Rigid flex PCBs were being fabricated satisfactorily with material
purchased from a qualified vendor. Subsequently, due to lower cost,
material was obtained from an alternate source which also met the
required MIL spec. Unfortunately, this resin matrix had a different
chemical formulation and could not be processed with the existing
shop laminating presses. However, before this fact was discovered
many manhours and expensive hardware were lost. This clearly illustrates the importance of thoroughly knowing your material. By adding chemical quality assurance specifications to our acceptance
requirements a user would be assured of the same prepreg formulation
with each lot of purchased material and that it would process the
same each time.

Prepreg suppliers have shown enthusiastic support of this approach and are receptive to working closely with customers to evolve tests which would assure consistent products. This type of cooperation with Lockheed's suppliers was a great help in solving our production problems. Therefore, to protect the proprietary nature of their formulations, the vendors and epoxy resins involved will not be specifically defined. The manufacturer of the acceptable prepreg will be referred to as Vendor A and the major epoxy resin used, Epoxy I. Vendor B will apply to the material which could not be processed and its epoxy resin, Epoxy II.

III. ANALYTICAL APPROACH

No-flow prepreg materials obtain their essentially "zero flow" characteristic by being highly B-staged or chemically advanced. This increased state of resin reaction makes processing of the PCB prepreg very difficult and exacting but has the advantage of being able to achieve complete cure without experiencing a fluid state during lamination. This is essential in maintaining registration of various layers during processing.

A preliminary explanation for what was happening in Lockheed's PCB production shop was that the two different prepreg materials had varying degrees of B-staging and/or they had different chemical formulations. A quick and easy analytical test to check this explanation was to obtain infrared spectra on both resins. These are shown in Figure 1. From this comparison the resin extracts appeared to be the same or at least very similar. These spectra, except for the peak at 2190 cm^{-1}, are the same as that of the higher molecular weight homologs of the diglycidyl ether of bisphenol A. The small peak at 2190 cm^{-1} is probably due to the curing agent, dicyandiamide. A logical conclusion from these infrared results (which later proved to be incorrect) was that the differences between the two prepregs was a simple variation in the amount of B-staging. An easy solution then would be for Vendor B to change his prepreg tower conditions to obtain a satisfactory material. This approach was tried without success.

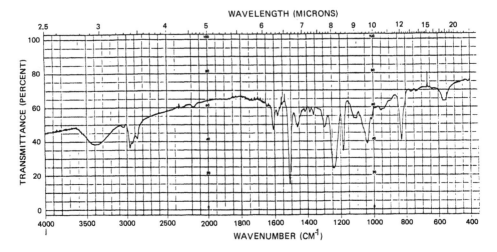

Vendor A Resin Extract

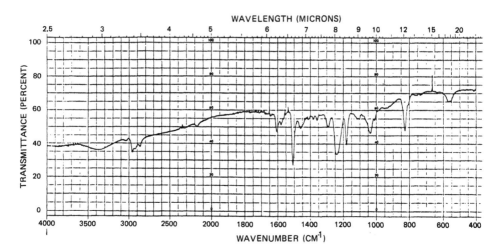

Vendor B Resin Extract

FIGURE 1.   Infrared spectra of no-flow resins.

Although other chemical characterization techniques (dynamic dielectric analysis, differential scanning calorimetry) gave useful information relating to the two materials' processing characteristics [9], liquid chromatography, specifically size exclusion or gel permeation chromatography (GPC), was the method which definitely confirmed the chemical differences between the resin matrices.

IV. EXPERIMENTAL

As previously discussed, no-flow prepregs are highly staged to obtain the proper flow characteristics. One complication arising from this advanced B-staging is the relative insolubility of these materials in organic solvents. To establish the solubility of the resins, the effects of several solvents and solvent mixtures were investigated. The outcome of this study is shown in Table 1. The best results were obtained with acetophenone. However, due to its unpleasant nature and probable incompatibility with chromatographic columns this solvent was eliminated as a potential candidate. Both tetrahydrofuran (THF) and chloroform are the easiest to work with and most popular for GPC separations so further examination was limited to the resin solutions involving these two solvents. Since centrifuging and basic filtration proved unsatisfactory in removing the high molecular weight gel, several attempts were made to consolidate the colloidal suspension so the remaining solution could be analyzed. Finally, a successful procedure was found. Relying on the opposite solubility effects of THF and toluene on the no-flow resins, it is possible to precipitate out the high molecular weight gel by adding a small amount of toluene to a THF-resin solution and vigorously mixing it on an electrical vortex mixer. This process is reversible as shown in Figure 2. The clear resin solution in "C" was easily filtered through a 0.45μ fluoromembrane filter for GPC analysis.

TABLE 1

Solubility of No-Flow Resins

Solvent	Observations
1. Methyl Ethyl Ketone	Slightly soluble, resin swells but remains attached to glass scrim cloth; solvent solution remains clear.
2. Toluene	Very slightly soluble; no swelling; solution clear.
3. 2-Ethoxyethyl Acetate (Ethyl Cellosolve Acetate)	Slightly soluble; no swelling – most resin appears gone from scrim; hazy solution.
4. 60/30/10 Cellosolve Acetate/MEK/Toluene	Same as 3.
5. Chloroform	Most resin gone from scrim; solution hazy but not as bad as 3 and 4.
6. Tetrahydrofuran	Most resin gone from scrim; solution hazy – clearer than 3 and 4 but not as good as 5.
7. Acetophenone	Most resin gone from scrim; solution slightly hazy.
8. Ethyl Acetate	Slightly soluble; resin swells but remains on scrim; clear solution.
9. Carbon Tetrachloride	Not soluble.
10. Chlorobenzene	Same as 9.
11. 50/50 Acetophenone/ Chloroform	Slightly soluble; hazy solution but not as bad as 100% 5.
12. 70/30 Acetophenone/ Chloroform	Slightly haze; no resin remains on scrim.
13. 80/20 Acetophenone/ Chloroform	Same as 12.
14. 1,1,1,3,3,3-Hexafluoro-2-propane	Same as 6.

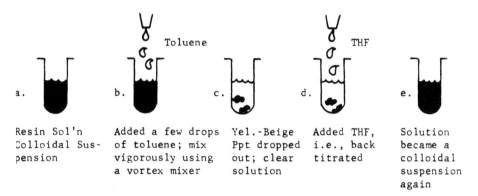

a.	b.	c.	d.	e.
Resin Sol'n Colloidal Suspension	Added a few drops of toluene; mix vigorously using a vortex mixer	Yel.-Beige Ppt dropped out; clear solution	Added THF, i.e., back titrated	Solution became a colloidal suspension again

FIGURE 2.   Solubility behavior of no-flow resin.

V.   RESULTS

A comparison of the chromatograms for the two no-flow resin extracts is shown in Figure 3. The differences are obvious and no question remained that these two materials were not chemically equivalent. By comparing these curves with those of commercially available epoxy resins a possible cause for the processing problems could be seen.

The matrix resin from the Vendor B prepreg is very similar to a commercially available material, Epoxy II, as shown in Figure 4. The Vendor A resin compares very closely with another but chemically different commercial resin, Epoxy I (Fig. 5). While Resin B is similar to a higher molecular weight homolog of the diglycidyl ether of bisphenol A (DEBPA), Resin A appears to have been formed by the in situ polymerization of the monomeric DEBPA with bisphenol A during base resin manufacture and the subsequent prepregging process. In view of this and the chromatogram comparisons the difference between the two materials appears to be one of molecular weight distribution. The chromatograms of the no-flow resins look the same as those of the prepreg starting resins because the high molecular weight gel was removed before analysis. The large peak at ~16.5 min in the Vendor A material is due to monomeric DEBPA, and this fraction of the matrix would be expected to gel more slowly. So even though the average

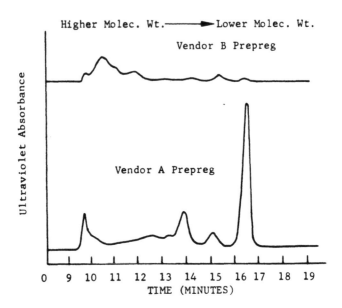

FIGURE 3.   Chromatograms of soluble portions of the no-flow resin matrices.

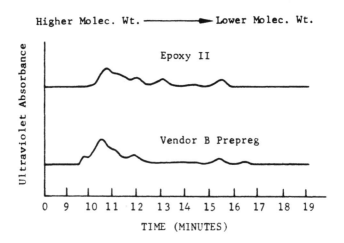

FIGURE 4.   Chromatogram of Vendor B resin vs. commercial epoxy resin II.

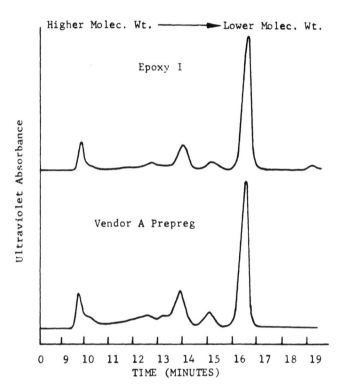

FIGURE 5.   Chromatogram of Vendor A resin vs. commercial epoxy resin I.

molecular weights of the two materials are similar, the presence of this low molecular fraction in the highly staged Resin A would be expected to produce superior laminating characteristics of the resultant prepreg. This, of course, was indeed the case.

Much more work needs to be done in the study of no-flow printed circuit board prepreg materials. Not only the critical nature of the PCBs themselves but the large investments associated with their fabrication make chemical characterization requirements mandatory in future purchase specifications.

VI. CONCLUSIONS

In order to improve the capability to predict ultimate performance
of raw materials and consistently produce high quality hardware,
composition specifications which guarantee that each lot of material
is chemically equivalent must be instituted. It is obvious from
this study that current MIL specifications are not adequate to define
raw materials used in the fabrication of laminated printed circuit
boards. The value of gel permeation chromatography for this purpose
has been illustrated and offers a fast and accurate quality assurance test to add to existing purchase specifications. A well developed quality control scheme including chemical characterization
tests will prove extremely cost effective by eliminating incoming
raw materials that would ultimately give poor parts as well as
classifying usable vs. unusable stock materials. Only by knowing
and understanding chemically the starting materials used can we hope
to be in total control of our fabricating processes.

REFERENCES

1. "Polymer Testing Saves Money in Electronics," Waters Associates,
    Inc. Technical Bulletin, Milford, Mass.

2. E. A. Eggers and J. S. Humphrey, Jr., Applications of Gel Permea-
    tion Chromatography in the Manufacture of Epoxy-Glass Printed
    Circuit Laminates, J. Chromatog., 55, 33-44 (1971).

3. D. K. Hadad, J. S. Fritzen, and C. A. May, "Exploratory Develop-
    ment of Chemical Quality Assurance and Composition of Epoxy
    Formulations," AFML-TR-76-112 and AFML-TR-77-217.

4. D. K. Hadad, Chemical Quality Assurance of Epoxy Resin Formula-
    tions by Gel Permeation, Liquid, and Thin Layer Chromatography,
    SAMPE Journal, 14, 4-10, July/Aug. 1978.

5. D. K. Hadad, "New Developments in Chromatographic Characteriza-
    tion and Quality Assurance of Composite Materials," American
    Institute of Chemical Engineers 72nd Annual Meeting, San Fran-
    cisco, Calif., Nov. 27, 1979.

6. C. A. May, D. K. Hadad, and C. E. Browning, "Physiochemical
    Quality Assurance Methods for Composite Matrix Resins," RP/C
    Proceedings, 33rd Annual Conference, Section 15-D, Washington,
    D.C., February 1978.

7. D. K. Hadad, "Chemical Quality Assurance Test Procedures for
    Advanced Composite Resin Matrices," Fifth Revision, May 1980 (to
    be released by Air Force Wright Aeronautical).

8. C. A. May, T. E. Helminiak, and H. A. Newey, "Chemical Characteri-
    zation Plan for Advanced Composite Prepregs," National SAMPE
    Technical Conference Series, 8, 274-294, Oct. 1976.

9. D. K. Hadad, A. Wereta, Jr., and C. A. May, "Chemical Characteri-
    zation of Highly Staged Circuit Board Material," 12th National
    SAMPE Technical Conference, October 1980, Seattle, Wash.

APPLICATION OF HPGPC AND HPLC TO CHARACTERIZE OLIGOMERS AND SMALL MOLECULES USED IN COATINGS SYSTEMS

C. Kuo
T. Provder
R. M. Holsworth
A. F. Kah

Glidden Coatings and Resins
Division of SCM Corporation
Strongsville, Ohio

I. INTRODUCTION

To comply with the governmental regulations on the volatile organic compounds (VOC) emission, new coatings systems such as High-Solids, Powders, Water-borne/Water-based, and radiation curable coatings have been developed. These new coatings systems either use water as the major solvent or contain low molecular weight polymers, oligomers, and reactive additives which when further reacted produce higher molecular weight and crosslinked polymers. The monitoring of the quality of raw materials and control of the oligomer/polymer content and molecular weight distribution (MWD) of the binders are critical to the coatings formulation.

In a previous paper [1], we have shown how high performance gel permeation chromatography (HPGPC) was used in (1) quality control of supplier raw materials, (2) guiding synthesis and processing, (3) modifying resin synthesis to improve end-use properties, and (4) correlating oligomer and polymer MWD with end-use properties. This paper will illustrate more HPGPC applications. The use of high performance liquid chromatography (HPLC) for high resolution analysis of oligomers and small molecules used in the above-mentioned environmentally acceptable coatings also will be discussed.

II.   EXPERIMENTAL

A.   High Performance Liquid Chromatography (HPLC)

The HPLC instrumentation used in this study consisted of a two pump
Varian model 8500 liquid chromatograph equipped with solvent gradient
programming capability via the Varian multilinear solvent programmer.
The detector used with the instrument was a Variscan UV-Visible
continuously scannable spectrophotometer with a 200-800 nm wavelength range equipped with 8 μl cells.   The types of columns used in
this study were Varian Instruments Silica, MicroPak SI-10 and the
bonded phase columns MicroPak CH-10, MCH-10, CN-10 and NH_2-10.   Also
used were Waters Associates μ-Bondapak-C_{18} and Hewlett-Packard RP-8.
The mobile phase solvents used in this study, except water, were
Burdick and Jackson distilled in glass solvents.   Water was purified
by passing deionized water through Millipore's Milli-Q System which
consisted of one carbon cartridge and two ion-exchange cartridges.

B.   High Performance Gel Permeation Chromatography (HPGPC)

The HPGPC used was the Waters Associates model 150C ALC/GPC which is
capable of being operated at elevated temperatures.   In this study,
the instrument was operated at 40°C with Burdick and Jackson distilled in glass THF as the eluting solvent.   The sample column bank
consisted of two 50 cm Varian Instruments MicroPak TSK Gel Columns
(TSK 2000H and TSK 3000H).   The flow rate was set at 1.0 ml/min and
the recorder chart speed was set at 1.0 cm/min.

The column plate count was determined from the expression

$$\text{Plate count} = 16 \ (V_R/W_b)^2 \qquad\qquad (1)$$

where V_R is the retention volume and W_b is the baseline width of the
plate count standard.   Using o-dichlorobenzene as the plate count
standard yielded 23,000 plates for 100 cm of column.

The specific resolution of the column set was determined from
the expression derived by Bly [2]:

$$R_s = \frac{2(V_{R_2} - V_{R_1})}{W_{b_1} + W_{b_2}} \cdot \frac{1}{\log_{10}(M_1/M_2)} \tag{2}$$

where V_{R_2} and V_{R_1} are retention volumes, W_{b_1} and W_{b_2} are baseline widths, and M_1 and M_2 are peak molecular weights for polymer standards 1 and 2, respectively. The R_s values of this column set determined from pairs of polystyrene standards with various molecular weights are listed in Table 1 and a molecular weight calibration curve is shown in Figure 1. It is interesting to note that the specific resolution obtained for this column set at 1.0 ml/min from the two polystyrene standards ($\overline{M}_w$ = 37,000 and $\overline{M}_w$ = 2,100) is the same as that of the μ-Styragel* column set described previously [1]. The flow rate for that system was set at 0.6 ml/min, and the column length was 180 cm. It is also noted that the plate counts are equivalent for these two sets at the different flow rate conditions.

A series of normal alkanes was run with the Model 150C. Their molecular weight, density at 20°C and the calculated bulk molar volume are listed in Table 2. The specific resolutions, R_s, for various pairs of alkanes are listed in Table 3. Krishen and Tucker [3] had shown some higher values for their selected pairs with a 60 cm TSK 2000H column at a flow rate of 0.5 ml/min.

TABLE 1

Specific Resolution for Various Polystyrene Pairs
with Different Molecular Weights[a]

Nominal molecular weights	2100	4000	10,000
4,000	1.58		
10,000	2.19	2.79	
37,000	2.18	2.49	1.91

[a]Polystyrene standards obtained from Pressure Chemical Co., Pittsburgh, Pa.

---

*Styragel is a Registered Trademark of Waters Associates, Milford, Mass.

TABLE 2

Physical Properties of Normal Alkanes

Sample	Carbon number	Molecular weight (M)	Density (ρ) at 20°C	Molar volume (M/ρ) ml/mol
n-Undecane	C_{11}	156.32	0.74017	211.2
n-Dodecane	C_{12}	170.34	0.7487	227.5
n-Tridecane	C_{13}	184.37	0.7564	243.7
n-Pentadecane	C_{15}	212.42	0.7685	276.4
n-Hexadecane	C_{16}	226.45	0.7733	292.8
n-Octadecane	C_{18}	254.51	0.7768	327.6
n-Eicosane	C_{20}	282.56	0.7886	358.3
n-Tetracosane	C_{24}	338.67	0.7991	423.8
n-Dotriacontane	C_{32}	450.89	0.8124	555.0
n-Pentatriacontane	C_{35}	492.97	0.8157	604.4

TABLE 3

Specific Resolutions for Various Pairs of N-Alkanes

	C_{11}	C_{13}	C_{15}	C_{16}	C_{18}	C_{20}	C_{24}	C_{32}
C_{13}	11.77							
C_{15}	12.90	12.17						
C_{16}	12.20	11.65	13.30					
C_{18}	12.00	11.39	12.58	11.33				
C_{20}	11.78	11.2	12.38	11.38	11.73			
C_{24}	12.83	12.22	13.66	12.61	12.92	12.79		
C_{32}	12.01	11.47	12.71	11.80	11.96	11.73	12.46	
C_{35}	10.65	10.10	11.00	10.2	10.19	9.88	9.84	4.10

The calibration curves generated by plotting the molar volume and molecular weight against the retention volume are both linear as shown in Figure 2. Also shown on the graph are three data points for the cyclic alkanes. They eluted later than their linear counterparts, as expected. For comparison purposes, a linear calibration curve also is shown for the polystyrene oligomers.

III.   RESULTS AND DISCUSSION

A. High Performance Gel Permeation Chromatography (HPGPC)

The HPGPC applications of a number of low molecular weight resins used in powder, high solids, water-borne and radiation curable coatings have been shown in the previous paper [1]. Also included in that paper was a brief description of each coatings technology. In the following, additional HPGPC applications will be shown.

High Solids Coatings. The polyesters used in high solids coatings are low molecular weight oligomers. Generally, they are synthesized by condensation reaction from dicarboxylic acids (or anhydrides) with diols. On occasion a third component, such as olefin oxide, also is added. The resulting molecular weight and distribution are a direct result of the molar ratio of di-acid/olefin oxide to diol in the reactor. Figure 3 shows the HPGPC traces of three polyesters made from various molar ratios of anhydride/olefin oxide to diol. It is seen that the molecular weight of the resin increases as the molar ratio of anhydride/olefin oxide to diol increases, as expected.

Figure 4 shows the HPGPC traces of two polyester resin samples. One sample had good "shelf stability" while the other sample was unstable on standing over a two-week period. It is seen from the chromatograms that the component which eluted at retention volume ∿30.5 ml was present in excessive amounts for the problem sample. This component apparently crystallized on standing and caused haze and then precipitated. Identification of the presence and amount

of this component will help resin chemists to control and eliminate
the instability problem.

   For the high solids polyesters to be useful as a protective
coating, the low molecular weight resin has to be cured to form a
crosslinked film. In order to compensate for the decrease in molecular weight of a polymer designed for high solids coatings, there
is an increasing dependence on the crosslinking agent, for the
development of mechanical properties. It becomes important to care-
fully match the crosslinking agent with the polymer both in terms of
reactive functionality and MWD. The most commonly used crosslinking
agents for high solids polyesters are melamine resins.

   Figure 5 shows the HPGPC traces of some of the melamine crosslinkers. M-1 is highly methylated and is claimed to be monomeric,
though at least four components are obviously present. M-2 is a
partially methylated resin. It is claimed as polymeric, which is
evidenced by the higher content of components in the higher molecular weight region. Due to the fact that it is only partially
methylated it has a higher tendency toward self-condensation. This
phenomenon is demonstrated by comparing the MWD between the new and
old M-2 resins as shown in the chromatograms. M-3 is a butylated
resin and is also claimed as monomeric. This material also has a
high tendency for self-condensation, presumably because of the
steric hindrance of the bulky butyl group which interferes with further alkylation. The level of high molecular weight could be a
direct reflection of the self-condensation reaction which would im-
part less impact resistance to a cured coating assuming all other
parameters are the same.

Powder Coatings. In powder coatings, one of the most frequently used
curing agents is blocked isocyanate crosslinkers. In the synthesis
reaction, the level of moisture present in the co-reactants will affect the MWD and properties of the resulting crosslinker. This is
due to the high reactivity of the $N=C=O$ functionality. Figure 6
shows the chromatograms of three isocyanate crosslinkers made with

different amounts of water present in the reactor. The one with 0%
water was a control. The other two samples were made with 0.5% and
2% of water being deliberately added as a coreactant. The weight
percent was based on the weight of one of the major coreactants.
The absolute amounts of water are about one-third of the amounts
shown in Figure 6. It is seen from the chromatograms that the
molecular weight distribution of the isocyanate crosslinkers made
in the presence of water is different from that of the control sample. In addition to the building up of the molecular weight, the
level of the component eluted at retention volume ∿27 ml is increasing with the amount of water added. A previous study [1] showed
that the presence of this component in excessive amount was one of
the reasons why some of the isocyanate crosslinkers were overly reactive.

Radiation Curable Coatings. Among the radiation curable coatings
systems, i.e., via ultraviolet (UV), electron beam (EB) and Infrared
radiation, UV curable coatings systems are the most widely used today. The formulation for UV coatings generally consists of multifunctional oligomers, reactive diluents, a photosensitizer and may contain
pigments. Figure 7 shows the oligomer distribution of a typical oligomer used in UV coatings for floor tiles application. The proper
level of functionality and molecular weight range is critical for the
development of ultimate mechanical properties in the resulting coating. In an EB curable system, unsaturated oligomers were mixed with
reactive monomers. Free radicals are generated on reactive functionalities upon exposure to the electron beams. There is no photo-
initiator needed. Figure 8 shows the HPGPC trace of a commercially
available electron beam coating system. Infrared spectroscopy
analysis revealed that the major component appeared to be a mixture
of some glycol diacrylate and acrylic oligomers. The molecular
weight of the oligomer is very low. This can be seen from Figure 8
by comparing the oligomer distribution with that of the 600 molecular
weight polystyrene standard.

Quality of Reactant. Hydrogenated bisphenol-A (HBPA) is another
type of material finding increasing usage in certain coating systems. Today, the number of suppliers is quite limited and the
quality varies; not only from supplier to supplier, but also from
lot to lot with the same supplier. Figure 9 shows the HPGPC
chromatograms of HBPA from three suppliers. The level of impurity
is detectable. It is found that the minute variation in impurity
makes a significant difference in coatings performance. Therefore,
if the HBPA from supplier F is substituted for the HBPA from supplier J, a readjustment in formulation has to be made. It is found
that the impurity is not the unhydrogenated BPA, the precursor, because BPA eluted earlier than HBPA. In this case, the aromatic
rings appear to be bulkier than the cyclohexane rings. To confirm
this idea, cyclohexanol and phenol were analyzed under the same
conditions. The cyclohexanol eluted at $\sim$32.6 ml which is later than
phenol at $\sim$32.2 ml.

B. High Performance Liquid Chromatography (HPLC)

Nonionic Surfactants. Some nonionic surfactants were investigated
by normal phase liquid chromatography (LC) for the immediate purpose
of qualitatively fingerprinting the surfactants for eventual quantitative pattern recognition analysis. The nonionic surfactants are
alkyl aryl polyether alcohols or alkyl phenyl ethers of polyethylene
glycol and have the general chemical formulas shown in Table 4.
Structure 1 is the octylphenol hydrophobe of the reaction of
octylphenol with ethylene oxide (EtO) and Structure 2 is the nonylphenol hydrophobe of the reaction of nonylphenol with EtO. As the
number of EtO units increases the surfactant becomes more water soluble. Therefore, a surfactant material containing one EtO unit
would be more hydrophobic or less polar than a surfactant material
containing 10 EtO units. However, as will be shown, these surfac-

TABLE 4

Nonionic Surfactant Structure Description

Str. 1 $C_8H_{17} - \hexagon - (OCH_2CH_2)_x OH$ Octylphenol series	Str. 2 $C_9H_{19} - \hexagon - (OCH_2CH_2)_x OH$ Nonylphenol series	Nominal ethylene oxide units
Triton X-15	--	1
Triton X-45	Triton N-57	5
Triton X-114	--	7-8
Triton X-100	Triton N-101	9-10
	Igepal CO-430	4
	Igepal CO-630	9

tant materials are comprised of a distribution of oligomers of varying EtO content. The surfactants investigated are shown in Table 4.

Figure 10 shows the experimental conditions that were established for nonionic surfactant analysis after considerable
experimentation to cover the range of material shown in Table 4.
The column best suited for this analysis is the MicroPak CN-10 column.
The solvent system gradient composition goes from hexane doped with
10% isopropyl alcohol (IPA) to IPA doped with 10% water. The water
is necessary to flush the column of all water soluble materials and
the IPA "couples" the water/hexane extremes. The nonlinear gradient
was established to provide optimum peak resolution.

Figure 11 shows Triton X-15 run under those conditions. The
nonlinear gradient is displayed on all chromatograms, as well as the
time scale for analysis for 12 mins. The sample is completely eluted
by 3 mins. Actually, this sample would not have required this
complex gradient for separation, but as stated previously, standard
conditions were desired for all samples. The major peak in Figure
11 can be assigned DP=1 or one EtO unit added. The second peak is
DP=2 and probably contains two added units of EtO. Figure 12 shows

the chromatogram for Triton X-45. This material should contain 5
EtO groups. Again, DP=1 as the first major peak through DP=6 or
DP=7 contributes to the bulk of the sample. Smaller amounts of
material are present up to DP=10 or DP=11. Figure 13 shows Triton
N-57 or the nonyl-equivalent of Triton X-45. Again, the major
portion of the sample is DP=7 and lower with a reasonably even
distribution. Although Triton X-45 and Triton N-57 give similar
chromatograms, there are significant and reproducible differences
which enable one to differentiate the two materials. Figure 14
shows the chromatogram for Triton X-114. The peaks appearing within the first 3 mins can be assigned to DP=1 through DP=3. The second section of the curve just under nine minutes contains peaks of
DP=9 to DP=10. Smaller amounts of peaks of DP=11 to DP=15 are then
eluted. The majority of the material seems to be between DP=4 and
DP=10, which would give the expected average of 7-8 EtO groups.
Figure 15 shows the chromatogram for Triton X-100. This sample contains only residual amounts of DP=1 through DP=3 oligomers with the
majority of the sample in the oligomer range between DP=4 and DP=15
to DP=16 oligomers. This would give the expected average value of
9-10 EtO groups. Figure 16 is the chromatogram of Triton N-101, the
nonyl-equivalent for Triton X-100. The oligomer distribution for
this sample is significantly different from that of Triton X-100.
The larger part of the sample would appear to be between DP=4 and
DP=10 with the major peak being DP=10. There also are present
minor amounts of DP=11 through DP=15 species. Igepal CO-430 which
contains 4 EtO groups, is shown in Figure 17. The oligomer distribution would appear to be evenly distributed from DP=1 through DP=4
and then tailing off from DP=5 through DP=10. Igepal CO-630 is
shown in Figure 18. This sample should contain 9 EtO groups. In
this case, the majority of the sample is from DP=8 to DP=10 with
smaller amounts of DP=1 through DP=15 present. Similar studies on
the LC separation of nonionic surfactants have been published by

other workers [4-7]. By quantitation of relative peak areas it
should be possible to develop pattern recognition techniques for
these nonionic surfactant materials.

Ionic Surfactants. Another type of surfactant frequently used in
preparing latices for water-based coatings is the linear alkylated
sulfonate compound. The material is anionic in nature when dissolved in water. Therefore, it is difficult to obtain a good separation with either normal phase LC method or the conventional
reverse phase LC method. Recently, paired-ion chromatography has
been used in our laboratory for the analysis of ionic surfactants.
The column used was a Varian Instrument MicroPak MCH-10 monolayer
reverse phase column. The mobile phase was CH_3CN/H_2O modified with
0.0025 M of tetrabutyl ammonium (TMA) phosphate buffer (PH = 6.9).
The gradient elution was from 25% CH_3CN to 50% CH_3CN as depicted in
the chromatograms. Figure 19 shows the chromatograms for two sodium
dodecyl phenyl sulfonate surfactant samples from different sources.
They behaved differently when used in the same emulsion polymerization reaction. As can be seen from the chromatograms, several
components are present in each sample. Although the LC conditions
are not yet optimized, the differences between the two surfactants
are readily seen. The paired-ion LC method for detecting the linear
alkyl benzene sulfonates in waste water was described by Gloor and
Johnson [8]. They achieved an optimum separation using
tetramethylammonium chloride (TMA) as the counterion. On the other
hand, Taylor and Nickless [9] modified the mobile phase with
cetyltrimethylammonium ($CTMA^+$) ions for the separation of partially
biodegraded linear alkyl benzene sulfonate.

Catalyst in Electrocoat Bath Permeate. In an electrocoating process
the metallic substrate is immersed into a bath of a low solids ($\sim$10%
NV) water-borne coating which is subjected to an electrical field.

The charged resins, pigments and additives are attracted to the
substrate, which functions as an electrode with opposite polarity,
and are discharged and deposited on the substrate.

Maintaining a constant chemical composition within a narrow
range of variation is necessary for the proper performance of the
bath and the cured coating which results from baking the electro-
deposited coating. One of the ingredients present in the electrocoat bath is the catalyst. These catalysts are used in the
electrocoat bath to promote the curing reaction between various
binder resins and melamine crosslinkers. The proper level of the
catalyst in the bath is critical for developing good performance
properties in the cured coating. One way to monitor the level of
the catalyst is by analyzing the electrocoat bath permeate. If the
level of catalyst detected in the permeate is too low compared to
the level present in the electrocoat bath, replenishment of the
catalyst might be necessary to maintain the chemical composition of
the electrocoat bath. The conventional analysis method in use mea-
sures the absorbance of the sample solution in a given wavelength
range with a UV spectrophotometer. In this method, the total absorbance of the permeate is measured which can include that of the
catalyst as well as that arising from impurities in the permeate.
Thus, it is susceptible to interference from other compounds.

We have used similar chromatographic conditions as used for
the anionic surfactants to analyze the catalyst in the electrocoat
bath permeate. Figure 20a shows the chromatogram of a clean permeate. The detector signal arises from the impurities in the bath.
The catalyst analyzed separately is shown in Figure 20b. It is seen
that the retention time of the catalyst is different from those of
the impurity peaks. Figure 20c is the chromatogram for the clean
sample after being spiked with the catalyst. The catalyst in the
electrocoat bath permeate is readily detectable and separable from
the impurities via the paired-ion chromatography LC method.

Plasticizers. A series of phthalate plasticizers, commonly found in
a variety of coatings systems, was separated into individual components

on a MicroPak CH-10 reverse phase column using a methanol/water
solvent system. A 50% to 100% MeOH gradient at various generation
rates was used at a flow rate of 2.0 ml/min, with the Variscan detector set at 233 nm. The chart speed is 0.5 in/min, and the detector sensitivity is at 0.5 AU full-scale. The phthalates
separated included: 1-Dimethylphthalate (DMP), 2-Diethylphthalate
(DEP), 3-Diphenylphthalate (DØP), 4-Dibutylphthalate (DBP), 5-
Diamylphthalate (DAMP), 6-Dioctylphthalate (DOP), and 7-Diisodecyl-
phthlate (DIDP).

Figure 21 shows the separation of the plasticizers at a 5%/min,
50 to 100% methanol (MeOH) gradient. There is a baseline separation
of the mixture and an analysis time of about 13 mins. At an 8%/min
gradient the analysis time is about 9 mins, with baseline separation
except for DØP and DBP. A 10%/min gradient gives the same results
with only a slight reduction in the analysis time. Figure 22 shows
a MeOH/water nonlinear gradient which results in baseline separation
and an analysis time of less than 10 min. The gradient starts at
65%/MeOH and is held for 1 min, increases at 10%/min for 1 min,
slows to 5%/min for 2 min and increases to 10%/min for 5 min or to
100% MeOH and completion of the analysis. The separation between
the diphenyl and dibutyl phthalates, although a baseline separation,
could still be improved.

Separation of four plasticizers, DMP, DEP, DBP, and DOP by
normal phase LC, is seen in Figure 23. Solvent conditions on the
silica, MicroPak SI-10 column are an isooctane/methylene chloride
(MeCl$_2$) (5% IPA), 90-10, isocratic solvent system. Other conditions
remain the same as those in Figure 21 except for flow rate. At the
previously used flow rate of 2 ml/min, the sample is separated to
baseline in less than 2 min. At 1 ml/min, the analysis time is
about 3 min and at 0.5 ml/min, analysis time is about 5 mins. Slowing the flow even more to 0.2 ml/min as seen in Figure 24, the
analysis time is about 6 mins. Slowing the flow to 0.1 ml/min gives
an analysis time of less than 10 min. The addition of two more
phthalates, DIDP and DØP to the sample, gives additional information

for these analysis conditions. The DØP is eluted on the tail of the
DBP between DBP and DEP, and the DIDP is eluted with DOP. Analysis
condition would have to be changed in order to analyze phthalates
higher than DOP.

UV Curable Coatings Components. 4,4'-Bis(diethylamino) benzophenone
is a photosensitizer used in UV-cured coatings. A knowledge of the
chemical purity of the material is necessary in order to have the
proper stoichiometry in a UV curable coatings formulation. The
column used was silica MicroPak SI-10. The solvent system was a
mixture of isooctane/MeCl$_2$ (5% IPA), 75/25, at a flow rate of 0.5
ml/min and the UV detector was set at 254 nm. As shown in Figure
25, a peak was eluted at 3 mins. Assuming efficient separation
with this system the material looks very pure. To verify the presence or lack of the possible starting material, diethylaniline, and
aniline and dimethylaniline were added to the sample and the sample
run under the same conditions. Three additional peaks for the three
added materials were eluted, proving that the 4,4'-bis(diethylamino)
benzophenone is quite pure.

Pentaerythritol triacrylate (PETA) is an oligomer used in UV-
cured coatings. Samples received from different suppliers gave
different cure behavior. Figure 26 shows the results of a GC analysis on two sources of PETA. Supplier A material consisted of tri-
and tetracrylates. This narrow distribution in the polymer resulted
in polymer crystallization making the material difficult to use.
Supplier B material consisted of di-, tri-, tetracrylates and higher
boilers plus peaks for acrylic acid and P-methoxyphenol (MEHQ)
inhibitor. A programmed temperature GC run to about 270°C took
about 28 min for complete analysis. Figure 27 shows the LC chromatograms for PETA from suppliers A and B. The silica MicroPak SI-10
columns were used with a solvent system of isooctane/MeCl$_2$, both
doped with 10% IPA at a 95/5 ratio. The flow rate was 0.5 ml/min
and the UV detector was set at 235 nm. The chromatograms for
samples A and B resemble the GC results in the distribution of

peaks. No attempt was made to trap the ultra-small amount of sample
which would elute from the analytical columns of the LC. By comparing GC and LC results, some peak designations were made. The time
for complete analysis is about 14 min. Running the sample at
different detector wavelengths can also help to identify some of the
peaks. An example of this is shown in Figure 28. Sample B was run
with the same conditions previously used except the solvent ratio
was 85/15. This caused the sample to elute much more rapidly for
an analysis time of about 8 mins. At 235 nm the sample appears as
before, only much sooner. At 290 nm the acrylate polymer does not
absorb in the UV. Therefore, it is not detected, but the MEHQ
inhibitor still absorbs strongly and is readily detected. Sample A,
run under the same conditions, shows no inhibitor present (Fig. 29).

Monomer Purity Analysis. N-(iso-Butoxymethyl) Acrylamide (NIBMA),
is a liquid crosslinking monomer which has been used in the manufacture of coatings and adhesives. This sample is not directly amen-
able to GC analysis without derivatization. A method for the HPLC
separation of the various components present in NIBMA has been
developed. The traces in Figure 30 show that at least eight well-
separated peaks were present. The peaks corresponding to NIBMA,
acrylamide and N-methylol acrylamide have been identified as indicated on the chromatograms. The column used was MicroPak NH_2-10.
The solvent system was a mixture of hexane/(10% IPA)/IPA (10% H_2O),
90/10, at a flow rate of 1 ml/min and the UV detector was set at
220 nm. The mobile phase was different from the one used in an
earlier stage of the analysis, where a gradient run of 5% to 50% of
dioxane in $MeCl_2$ was used as shown in Figure 31. The current operating conditions provided better resolution and better baseline
stability. LC work in this area reported in the literature has
emphasized the measurement of residual acrylamide monomer in
polyacrylamide [10-12].
    Bisphenol A (BPA) is a widely used raw material for making
epoxy resins. The purity of BPA is critical in the manufacture of

liquid epoxy as well as solid epoxy resins through the molecular
weight advancement process. The quality of the commercially available BPA varies from vendor to vendor. The success of a given epoxy
manufacturing process depends very much on the purity of the BPA.
Szap, Kesse and Klapp [13] had found the purity of commercial BPA in
the range between 93% to 99+% by an LC method. Figure 32 shows the
HPLC chromatograms of BPA from two suppliers. The variation in
purity is clearly demonstrated. These separations were done on a
MicroPak MCH-10 reverse phase column using CH_3CN/H_2O as mobile
phase. The gradient elution was run from 45% CH_3CN to 65% CH_3CN.
It is seen that the BPA from supplier A contains a fair amount of
impurities other than the major component, the 4,4'-isopropylidene
diphenol. The BPA from supplier B is very pure. If BPA from supplier A would be used in a polymerization process based upon the
purity of BPA received from supplier B, the final product would be
out of specification. A process reformulation would be needed.

High Resolution LC Analysis of Resin Oligomers. Two samples of
polyesters used in high solids coatings were also analyzed with HPLC.
Figure 33 shows the detailed separation of a polyester used in a
fabricated metal application. Also included in the figure is the MWD
of the same sample analyzed by HPGPC. The excellent resolution of
more than 20 peaks with HPLC is readily seen. In this HPLC separation the higher molecular weight species elute at increasing times.
The separation was done with a Varian monolayer MCH-10 column. The
solvent system was CH_3CN/H_2O with nonlinear gradient from 10% CH_3CN
to 90% CH_3CN holding at 10% CH_3CN for 2 mins; 6%/min increase for 5
mins; holding at 40% CH_3CN for 2 min; 3%/min increase for 10 min,
and finally, with 6%/min increase to 90% CH_3CN. The flow rate was
set at 2 ml/min and UV detector at 220 nm.

Figure 34 shows the HPLC chromatogram of another sample of high
solids polyester. This sample possessed a better salt spray resistance than the previous sample, presumably due to the higher level
of high molecular weight components as indicated from the HPGPC

traces. The HPLC chromatogram also shows a clear distinction between the two samples.

A solid epoxy resin was also analyzed with HPLC. Figure 35
shows the oligomer distribution analyzed with a Hewlett-Packard
RP-8 column. The solvent system was CH_3CN (50%):MeOH/H_2O:MeOH (10%)
with linear gradient from 60% CH_3CN (50%):MeOH to 100% CH_3CN (50%):
MeOH holding at 60% CH_3CN (50%):MeOH for 2 mins, 3%/min increase for
20 mins, and held at 0% for 5 min. The flow rate was set at 1 ml/
min and UV detector at 230 nm. An HPGPC chromatogram also is included for comparison.

The same sample was also analyzed with a Waters µ Bondapak C_{18}
column as shown in Figure 36. The solvent system was CH_3CN/H_2O with
linear gradient from 70% CH_3CN to 100% CH_3CN, holding at 70% CH_3CN
for 1 min, 3%/min increase for 10 min and hold at 100% CH_3CN for 10
mins. The flow rate was set at 1 ml/min and UV detector at 230 nm.
Although the solvent system and gradient are not exactly the same,
the better resolution of the C_{18} column at higher molecular weight
regions is apparent.

IV. SUMMARY

In this paper we have shown some applications of HPGPC and HPLC
for the analysis of coatings system components. The information
generated via the HPGPC and HPLC techniques have significantly
aided resin chemists and coatings formulators to tailor-make
environmentally acceptable coatings systems.

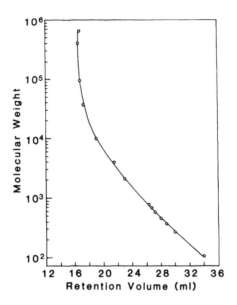

FIGURE 1. Polystyrene molecular weight calibration curve.

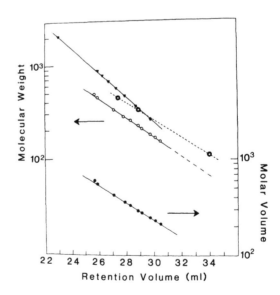

FIGURE 2. Calibration curves for n-alkanes (-o-), polystyrene oligomers (-▲-) and cyclic alkanes (···◉···).

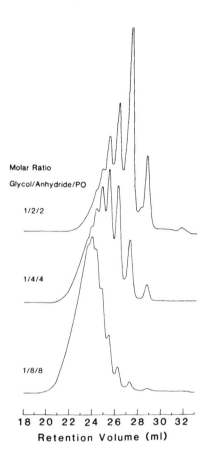

FIGURE 3.   HPGPC chromatograms of three model polyesters.

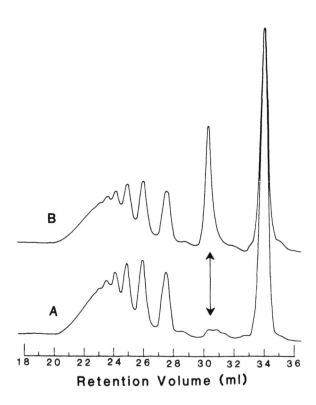

FIGURE 4.   HPGPC chromatograms of two high solids polyesters.

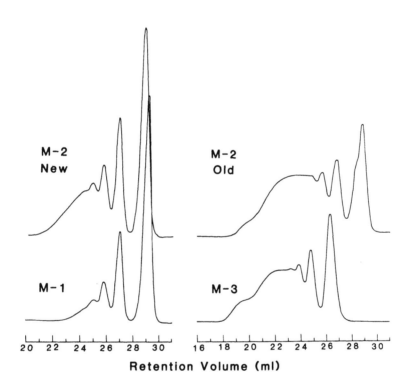

FIGURE 5.   HPGPC chromatograms of four melamine resins.

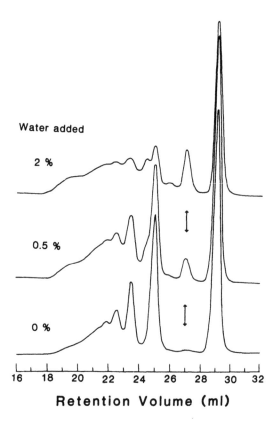

FIGURE 6.   HPGPC chromatograms of three blocked isocyanate crosslinkers.

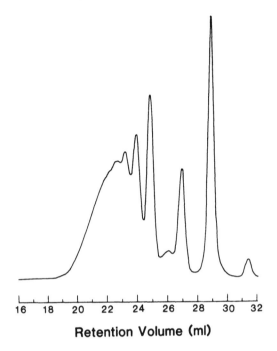

FIGURE 7. HPGPC chromatograms of a typical UV curable resin.

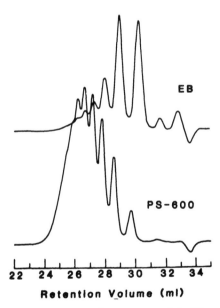

FIGURE 8. HPGPC chromatograms of a commercial EB curable coating and a polystyrene ($\overline{M}_w$ = 600) standard.

191

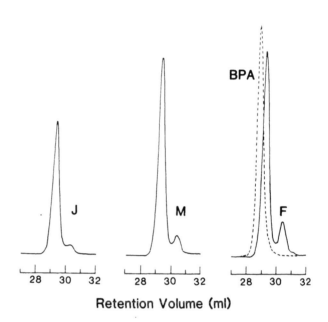

FIGURE 9.   HPGPC chromatograms of HBPA samples from different
suppliers.

SURFACTANT CONDITIONS

Column: CN-10
Solvent: Hexane (10% IPA) / IPA (10% H$_2$O)
Gradient: Non-Linear—0-100% IPA (10% H$_2$O)
          0 % —Hold— 1 min.
          2 %/min. — 5   "
          5 %/min. — 5   "
          8 %/min. — To 100%
Flow Rate: 0.5 ml/min.
Detector: 220nm.
Sensitivity: 0.5 A.U.
Chart Speed: 0.25 in./min.

FIGURE 10.   Liquid chromatograph conditions for nonionic surfactant
analysis.

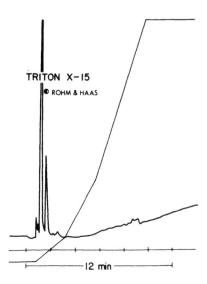

FIGURE 11. Liquid chromatogram for Triton X-15.

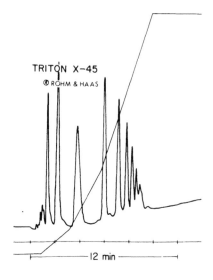

FIGURE 12. Liquid chromatogram for Triton X-45.

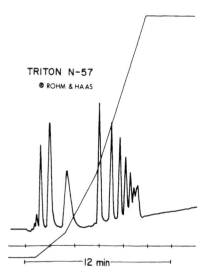

FIGURE 13.   Liquid chromatogram for Triton N-57

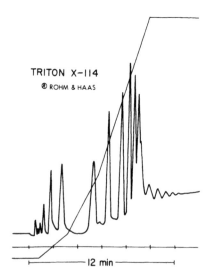

FIGURE 14.   Liquid chromatogram for Triton X-114.

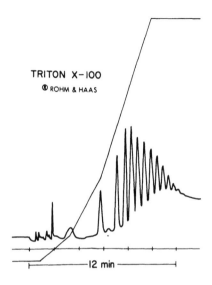

FIGURE 15.   Liquid chromatogram for Triton X-100.

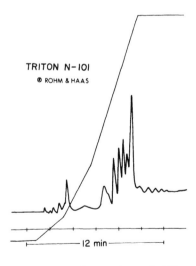

FIGURE 16.   Liquid chromatogram for Triton N-101.

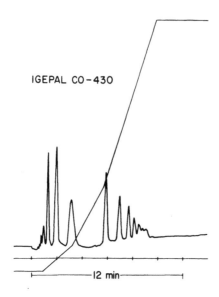

IGEPAL CO-430

FIGURE 17. Liquid chromatogram for Igepal CO-430.

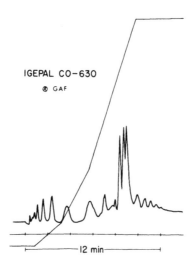

IGEPAL CO-630
® GAF

FIGURE 18. Liquid chromatogram for Igepal CO-630.

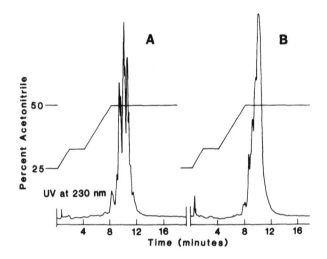

FIGURE 19.   Liquid chromatograms of two linear alkyl phenyl sulfonate samples.

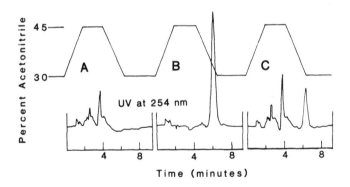

FIGURE 20.   Liquid chromatographic analysis of catalyst in electrocoat bath permeate.

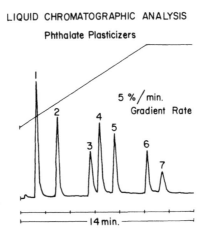

LIQUID CHROMATOGRAPHIC ANALYSIS

Phthalate Plasticizers

5 %/ min.
Gradient Rate

14 min.

FIGURE 21. Liquid chromatogram of the separation of seven phthalate plasticizers by reverse phase LC method.

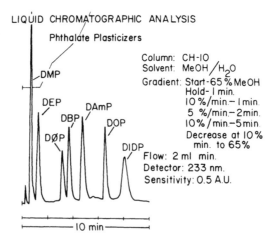

LIQUID CHROMATOGRAPHIC ANALYSIS

Phthalate Plasticizers

DMP

DEP   DAmP

DBP

DOP

DØP

DIDP

Column: CH-10
Solvent: MeOH/H_2O
Gradient: Start-65 % MeOH
          Hold- I min.
          10 %/min.- I min.
          5 %/min.-2min.
          10%/min.-5min.
          Decrease at 10%
               min. to 65%
Flow: 2 ml min.
Detector: 233 nm.
Sensitivity: 0.5 A.U.

10 min

FIGURE 22. Liquid chromatogram and conditions for analysis of phthalate plasticizers with a nonlinear gradient.

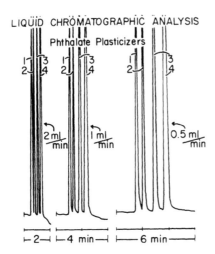

FIGURE 23.   Liquid chromatograms of the separation of four phthalate plasticizers at different flow rates.

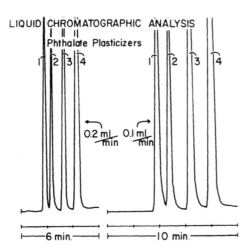

FIGURE 24.   Liquid chromatograms of the separation of four phthalate plasticizers at different flow rates (0.2 and 0.1 ml/min).

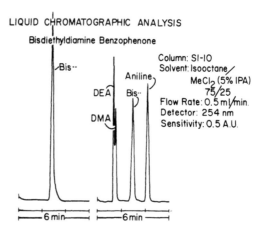

FIGURE 25. Liquid chromatogram of 4,4'-Bis(diethylamino) benzophenone and possible reaction products.

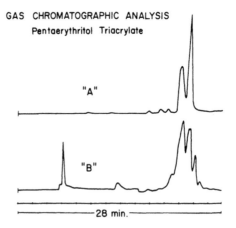

FIGURE 26. Gas chromatographic analysis of pentaerythritol triacrylate (PETA) from two suppliers.

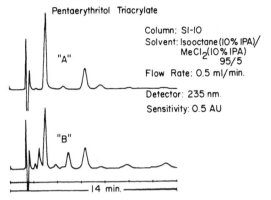

FIGURE 27.   Liquid chromatographic analysis of PETA from two suppliers.

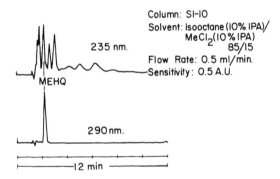

FIGURE 28.   LC analysis of PETA at two UV wavelengths.

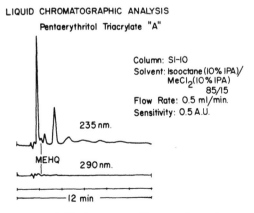

FIGURE 29.   LC analysis of PETA at two UV wavelengths.

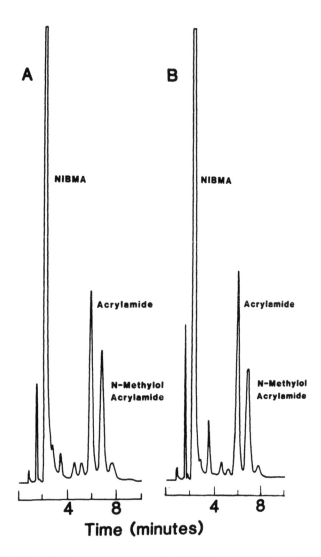

FIGURE 30. Liquid chromatograms of NIBMA from different lots.

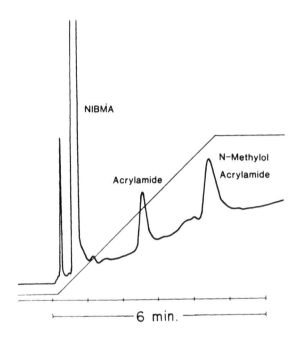

FIGURE 31.   Liquid chromatogram of NIBMA with a gradient elution.

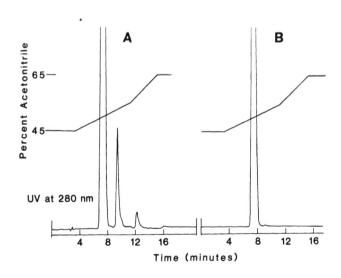

FIGURE 32.   Liquid chromatograms of BPA from two suppliers.

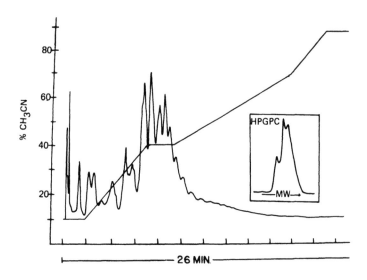

FIGURE 33. Liquid chromatogram of a polyester resins (lower molecular weight).

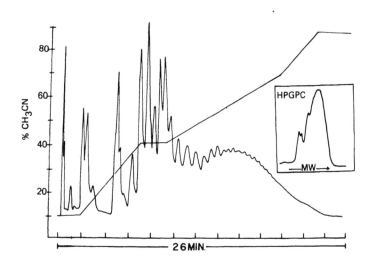

FIGURE 34. Liquid chromatogram of a polyester resin (higher molecular weight).

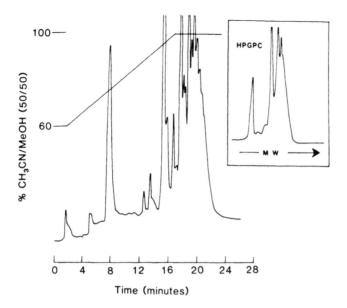

FIGURE 35.   LC analysis of a solid epoxy resin with a RP-8 column.

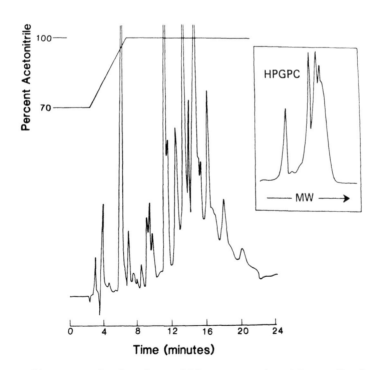

FIGURE 36.   LC analysis of a solid epoxy resin with a μ-Bondapak
C$_{18}$ column.

REFERENCES

1. C. Kuo and T. Provder, *ACS Symposium Series*, <u>138</u>, 207 (1980).

2. D. D. Bly, *J. Polymer Sci.*, Part C., <u>21</u>, 13 (1968).

3. A. Krishen and R. G. Tucker, *Anal. Chem.*, <u>49</u>, 898 (1977).

4. J. F. K. Huber, F. F. M. Kolder, and J. M. Miller, *Anal. Chem.*, <u>44</u>, 105 (1972).

5. K. Aitzetmuller, *J. Chromatog. Sci.*, <u>13</u>, 454 (1975).

6. C. F. Allen and L. J. Rice, *J. Chromatog.*, <u>110</u>, 151 (1975).

7. F. P. B. Van der Maeden, M. E. F. Biemond, and P. C. G. M. Janssen, *J. Chromatog.*, <u>149</u>, 539 (1978).

8. R. Gloor and E. L. Johnson, *J. Chromatog. Sci.*, <u>15</u>, 413 (1977).

9. P. W. Taylor and G. Nickless, *J. Chromatog.*, <u>178</u>, 259 (1979).

10. E. R. Husser, R. H. Stehl, D. R. Price, and R. A. Delap, *Anal. Chem.*, <u>49</u>, 154 (1977).

11. F. J. Ludwig, Sr., and M. F. Besand, *Anal. Chem.*, <u>50</u>, 185 (1978).

12. H. E. Skelly and E. R. Husser, *Anal. Chem.*, <u>50</u>, 1959 (1978).

13. P. Szap, I. Kesse, and J. Klapp, *J. Liquid Chromatog.*, <u>1</u>, 89 (1978).

ON LINE DETERMINATION BY LIGHT SCATTERING
OF MECHANICAL DEGRADATION IN THE GPC PROCESS

J. G. Rooney
G. Ver Strate

Elastomers Technology Division
Exxon Chemical Company
Linden, New Jersey

ABSTRACT

In addition to the effects of dispersion, adsorption,
concentration dependence of chain dimensions, etc. on elution volume
derived molecular weights, mechanochemical degradation of the polymer must be considered. Use of low angle light scattering, LALLS
(Chromatix KMX-6) as an on line detector, permits absolute M_w
determination across the chromatogram. Study of the flow rate and
initial polymer molecular weight dependence of the M_ws so determined
reveals that a significant portion of the apparent loss of resolution
at high molecular weights can be related to polymer degradation, for
typical micro particulate column packings (e.g., Waters E-Linear
Bondagel or Shodex 800 series). Above a given M_w for a given polymer
type, the whole polymer M_w becomes flow rate dependent down to at
least 0.3 cc/min (1,2,4 trichlorobenzene, 135°C, Waters 150C), with
M_w taken at a series of flow rates extrapolating to the statically
determined value. Typical M_w where degradation (at 0.5 cc/min) becomes severe is ca 7.0 × 10^5 for polyethene, whereas at $M_w \sim 1.5 \times 10^5$
10^5 no variation is seen up to 1.5 cc/min. Results are presented for
a variety of polymer types and MWDs. If the observed (degraded)
LALLS molecular weights are used in calibration plots, the apparent
flow rate dependence of elution volume is significantly reduced.

Comparison of the characteristic times for flow past packing
particles and maximum relaxation times for polymer coil rearrange-
ments suggests significant elastic effects should be present for
molecular weights of the order of 10^6. With Deborah numbers of order
unity it is found that the pressure drop across the columns exceeds
that expected on the basis of shear viscosity. Extensional flows are
most likely involved in the degradation process.

Since it is unlikely that mechanochemical degradation will
scale with hydrodynamic volume to more than a first approximation,
much molecular weight distribution data in the literature must be
spurious.

Such degradation is problematical for high speed MWD determination, but can be used as a tool to study the susceptibility of given
polymer structures to degradation. Such degradation studies can be
made quite informative by incorporating well-defined orifices in the
GPC instrument. Such defined geometries permit analysis of the flow
kinematics and establishment of strain rates needed to promote the
degradation immediately detected downstream by the LALLS apparatus.

I. INTRODUCTION

With the advent of on line molecular weight determination for gel
permeation chromatographic (GPC) separations [1-7], the process of
studying the effects of column type and operational variables [8-17]
on the GPC process has been simplified. Such molecular weight
determination also provides an opportunity to study effects whose
existence has been primarily speculative. It is common occurrence
that the degree of resolution loss at high molecular weights in a
particular column set is quite sensitive to flow rate and concentration. This sensitivity has been ascribed correctly, in part, to the
high concentration dependence of chain dimensions in good solvents,
and increased efficiency (plate count) at low flow rate and viscosity.

The possibility also must be considered that mechanochemical
degradation of polymer is occurring in the GPC apparatus [18]. On
line molecular weight detection provides an excellent opportunity to
study such an effect [51]. In addition, through use of appropriately
designed orifices, etc. placed either up or downstream of the columns,
the GPC apparatus itself can be used simultaneously to promote and
analyze mechanical degradation.

Presented here is a brief analysis of column characteristics
with regard to deformation rates, the relevance of these rates to
those known to produce degradation, and finally experimental data
demonstrating degradation for two column and several polymer types,
molecular weights and distributions.

II. EXPERIMENTAL AND RESULTS

Instruments used were a Waters 150C GPC (Waters Associates, Milford, Mass.), with a Chromatix KMX-6 (Chromatix Inc., Sunnyvale, Calif.) light scattering photometer (He-Ne 633 nm), either used statically or in line between the GPC columns and differential refractometer. All data reported here were collected using as solvent 1,2,4 trichlorobenzene (TCB, Aldrich) filtered under N_2 to 0.025 µm after being inhibited with 0.05 wt. % Ionol, a hindered phenol. Solutions were prepared by weight using 1.315 g/cc for the density of TCB at 135°C [19]. Polymer and solvent were heated in an oven in sealed bottles to 120°C and were dissolved with gentle swirling. If not used immediately, solutions were stored at ambient temperature in the dark. In ancillary studies antioxidant concentration was varied over a tenfold range with no observable variation in oxidative degradation, i.e., none, occurring from ca 0.02 to 0.15 wt. % inhibitor. There is no question that oxidative degradation can occur in unprotected solutions of the hydrocarbon polymers studied. The effects of sunlight (even on TCB alone) are enormous. We feel certain that the effects reported herein are not due to variation in oxidative rather than mechanochemical phenomena. All solutions studied were freshly prepared, although storage at 23°C in the dark for periods of days did not appear to influence the results. Certain ethene propene copolymers and polyethylenes withstand at least 16 hours at 135°C in solution without significant oxidative degradation.

For static scattering a range of concentrations was examined, each individually prepared and filtered (0.2 µm α-8 Metricel, Gelman) directly into the scattering cell which cell was also used for the on line GPC work. No filtration was employed with the GPC (when columns were in place) other than the 2-5 µm in-line precolumn filters, and it appeared that the GPC columns themselves were at least as effective as the 0.2 µm filtration. A series of experiments were performed with no columns in place. To reduce scattering due to dust a 0.45 µm filter (α-6 Metricel, Gelman) was located in place of the

columns (Chromatix filter holder design). The columns were more
effective in eliminating particulates. Static scattering data were
analyzed as $(Kc/\overline{R}_\theta)^{1/2}$ versus c where $K = 2\pi^2 n^2 (dn/dc)^2 (1 + \cos^2 \theta)/\lambda^4 N$
where n is refractive index, λ is wavelength, N is Avogadro's number,
dn/dc is the specific refractive index increment, θ is the scattering
angle ($\sim 5°$) and $\overline{R}_\theta$ is the excess scattering of polymer solution over
solvent. For our case this simplifies to $K = 2.04 \times 10^{-8} n^2 (dn/dc)^2$.
The Rayleigh ratio for TCB at 135°C is 3.5×10^{-5} cm^{-1}. The KMX-6 was
operated at a gain such that TCB scattering (photomultiplier output)
was 1280 mv with the 6-7° annulus and 0.2 field stop.

Effective scattering angle is 4-5° and no extrapolation to zero
angle is needed. Optical parameters are given in Table 1. dn/dc was
determined using the Chromatix KMX-16 differential refractometer.
Data were analyzed using Chromatix's software MOLWT·MAC, which in-
cludes despiking and baseline routines and second virial coefficient,
A_2, corrections. A_2 was evaluated statically over a range of $\overline{M}_w$ from
10^3 to 10^6 for the ethene propene polymers. The corrections are
small. Approximate relationships used for the other polymers appear
in Table 1. In this program a volume for the interdetector holdup
must be supplied, but no provision is made for tailing [20] or other
flow imperfections. Use of an inaccurate value for the holdup can
result in large errors in the $\overline{M}_w$ determined in the tails of the
distribution, however overall $\overline{M}_w$ are not as significantly affected,
e.g., ±10% for any reasonable estimate of the holdup.

Operation of our apparatus with columns removed produces traces
such as those in Figure 1. Certain features are apparent. First,
elution time to onset of a signal at the DRI indicates a piping and
KMX-6 cell dead volume of ca. 0.4 cc. This is not significantly altered by the presence of the filter. This is larger than anticipated
on the basis of pipe volume (0.08 cc) and quoted KMX-6 cell volume
(~ 0.055 cc). The injected plug of solution obviously does not have
square wave character. The onset takes ca. 0.1 cc and tailing at
least 0.2 cc. The pattern is not symmetrical. This extra column
dispersion should be quite significant for the low volume E-Linear
columns. Since the volume exceeded that calculated from pipe and

TABLE 1

Whole Polymer $\overline{M}_w$[a]

Polymer type	Sample name		$\overline{M}_w/\overline{M}_n$	Static $\overline{M}_w$[b]	Flow rate dependence of $\overline{M}_w$ E-Linear/Shodex (flow rate ml/min)			
					0.3	0.5	1.0[d]	1.4[d]
Poly coethene propene (0.03%)	3354C	-0.104	~2.2	1.1×10^6	0.87×10^6	0.82×10^6	$0.69 \times 10^6/0.73 \times 10^6$	$0.69 \times 10^6/0.56 \times 10^6$
	3208D	-0.104	2.2	4.4×10^6	--	4.3	--	--
	3208C	-0.104	2.1	3.1×10^5	--	3.3×10^5	--	--
	JBG #11	-0.104	2.1	1.6×10^5	$1.6 \times 10^5/1.6 \times 10^5$	$1.6 \times 10^5/1.6 \times 10^5$	$1.6 \times 10^5/1.6 \times 10^5$	$1.6 \times 10^5/1.6 \times 10^5$
Polystyrene (0.05%)	Duke 500	0.046	~1.3	(7.0×10^6)		$/6.6 \times 10^6$	$/4.2 \times 10^6$	$/2.6 \times 10^6$
	Duke 501	0.046	1.2	4.1×10^6	$4.0{-}4.6 \times 10^6/4.0 \times 10^6$	2.6×10^6		$2.1 \times 10^6/2.6 \times 10^6$
	Pressure Chem 13a	0.046	1.1	6.7×10^5	--	6.5×10^5	$/(5.4 \times 10^6)$	
Poly(octene-1) A		(-0.10)	1.1	6.6×10^6	--	--	--	1.8×10^6

[a]Midrange concentration; see text.

[b]Virial coeff. α, β if $A_2 = \beta M^{-\alpha}$ are; poly co (ethene–propene), 0.185, 9.9×10^{-3}; polystyrene, 0.2, 5×10^{-3}; polyoctene-1 approximated as for poly co (ethene propene).

[c]n for TCB @ 633 nm 135°C is 1.502 (Chromatix data sheets).

[d]These values probably represent upper bounds due to the excess scattering at high flow rates as discussed in the text.

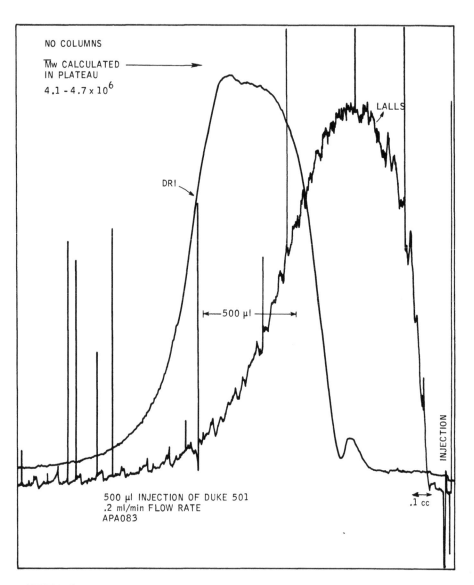

FIGURE 1.   Degradation absent without columns.

cell dimensions, we experimentally determined the portion of the piping volume which lies between the KMX-6 cell and the DRI by two methods, both of which no doubt included some of the spreading effects. A useful discussion has recently appeared [52].

During normal operations, for all but monodisperse polymers, peaks in the DRI and light scattering chromatograms should not coincide on the elution time axis. However, they will coincide if there is no fractionation. If there is "no" fractionation, however, low molecular weight "impurities" will elute at the same time as the polymer. Although a systematic shift in the total DRI trace is normalized out, any differences in adsorption characteristics of the polymer and impurities could lead to variations across the "chromatograms." If this complication is ignored and samples are injected with no columns present the elution volume between DRI and LALLS peaks is ca. 0.25 cc. It is apparent, however, that there is tailing or adsorption because constant $\overline{M}_w$ cannot be obtained across the entire chromatogram. Furthermore, for very high molecular weight polymer the chromatograms assume irregular shapes.

A second method is as follows. Replicate analyses are performed on an extremely narrow molecular weight distribution sample using columns with low total interstitial volume (E-Linears). The molecular weight should be in a range where the columns are performing effectively. For our system a "polyethylene" prepared by hydrogenating anionically prepared polybutadiene proved suitable. $\overline{M}_w$ was 1.0×10^5 and $\overline{M}_w/\overline{M}_n \sim 1.13$. NBS 1484 is somewhat broader and therefore less suitable. Polystyrenes exhibit tailing on our inorganic columns (Bondagel, Waters), the columns in place for the holdup study. Data are collected with a series of values assigned to the interdetector delay (or the software can be modified to permit variation of the delay on a single run). Examination of the $\overline{M}_w$ across the chromatogram reveals, at high $\overline{M}_w$, decreasing values for $\overline{M}_w$ at a given elution time as the interdetector holdup is varied mathematically, from small to large values. The reverse is true at low molecular weight. At extreme values for the delay, $\overline{M}_w$ calculated

will show "reversion." This sets an upper bound for the volume.
When the refractometer data alone are analyzed to give $\overline{M}_w/\overline{M}_n$ based
on an elution time calibration, an upper bound for this ratio on
the sample can be obtained, if resolution is good and dispersion
corrections are not made. $\overline{M}_w/\overline{M}_n$ calculated from the light scattering output monotonously decreases as the delay is increased. By the
nature of the averaging process it is necessary that $\overline{M}_n$ from the
LALLS data be larger than the true value, therefore a delay must be
used which is large enough so that $(\overline{M}_w/\overline{M}_n)_{LALLS} < (\overline{M}_w/\overline{M}_n)_{GPC}$. This
provides a lower bound. For a set of three E-Linear columns in our
system we find that the delay time is fixed at 0.19 cc to within a
few percent by these two bounds. This value was used throughout
the following analysis.

Two sets of columns were used. One consisted of three E-Linear
columns (Waters) which are packed with 0.1-15 μm inorganic "porous"
irregular fragments coated with a polyether layer to reduce adsorption. The second set were styrene divinyl benzene gels (Shodex;
marketed in the U.S. by Perkin Elmer). Columns designated 805, 802,
804 and 803 were employed in that order. The styrene gels were
spherical and ca. 5-10 μm. We sized packings by electron microscopy
after use.

Calibration of the elution time-molecular weight relationship
was performed using a variety of narrow MWD polyethylenes,
hydrogenated polybutadienes, hydrogenated polyisoprenes and polystyrenes. Data appear in Figure 2. Both data and selected points
from the calculated calibration curves are shown for the Shodex
columns. The difficulty in fitting, with quadratic or cubic equations, what appears to be a smooth curve through the data is to be
noted. The usual myriad of concentration and flow rate effects were
observed. In particular, for the E-Linear columns, elution volumes
were found to decrease with decreasing flow rate and decreasing
concentration. (See discussion in connection with Figure 5 below.)
For the Shodex columns, elution volume of the peak molecular weight
was less sensitive to flow rate than for the E-Linears (however,

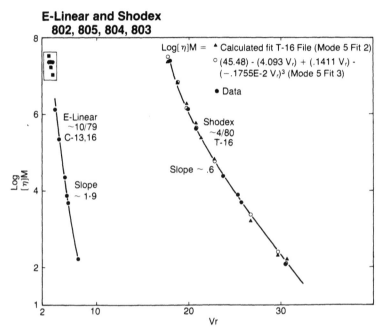

FIGURE 2.   Typical calibration curves.   Strong flow rate and
concentration dependence at large [η] M--these data at 0.5 cc/min
except for highest [η] M points on E-linears.

TABLE 2

Calibration Blends

Blend #	Polymer type	Component	Log [η]·M	$\overline{Mw}$
1	polyethene & hydrogenated polybutadiene	HPB 4.018	6.11	3.6×10^{5}
		NBS 1484	5.35	1.1×10^{5}
		NBS 1483	4.36	3.2×10^{4}
		NBS 1482	3.7	1.3×10^{4}
2	polystyrenes	Duke 501	7.22-7.37	4.1×10^{6}
		Pressure chem 3b	5.35-5.63	4.1×10^{6}
		Pressure chem 7b	3.83-3.88	3.7×10^{4}
		Pressure chem 11b	2.23	4.0×10^{3}

Polyethylene $[η] = 3.92 \times 10^{-4} M^{0.725}$ [19].

Polyco(ethene propene) $[η] = 2.92 \times 10^{-4} M^{0.726}$

Polystyrene $[η] = 1.9 \times 10^{-4} M^{0.655}$ [20] or

$[η] = 1.21 \times 10^{-4} M^{0.707}$ [21].

degradation was as severe). Such performance was generally studied
in four component blends as shown in Table 2 and Figure 3, although
individual polymers, especially Duke 501, 500 and 3354C, were also
studied. The sum of the component concentrations in the blends was
approximately the nominal values indicated below. Hydrodynamic
volumes calculated for polyethylenes using either the NBS values
[21] for [η] or published Mark Houwink constants [22] gave results
"consistent" with those calculated for polystyrenes using either of
two published [23,24] sets of Mark Houwink parameters.

Although elution volumes varied significantly with flow rate,
for high molecular weight samples, the peak shapes were amazingly
insensitive. No "bimodal" or other irregularities were observed
(Fig. 4). Similarly, if MWD parameters were calculated no systematic
changes were observed for typical samples (Fig. 5). If the peak
molecular weight, or the total $\bar{M}_w$, calculated from the light scattering were used to calculate an actual hydrodynamic volume the "flow
rate dependence" observed in Figure 2 (see boxed areas, horizontally
placed circles at log [η] · M ∿ 7.2) becomes a normal extension of
the calibration curve (filled squares). Apparently dispersion effects are dominant in determining the peak shapes as the limits of
resolution are approached. Furthermore, degradation occurs early
enough in the columns so that the resultant elution volume correlates
somewhat with the actual degraded molecular weight.

Total integrated areas of DRI versus elution volume traces were
independent of flow rate indicating an absence of adsorption effects.
That $\bar{M}_w$ was independent of flow rate for low molecular weight polymers indicates the absence of flow rate irregularities.

For the studies of degradation nominal injection values were
as follows. For the three E-linear columns the product of injection
volume concentration and $\bar{M}_w$ was constant at ca 27 g^2/mole (e.g.,
120 μl of 0.15% solution at 1.5×10^5 $\bar{M}_w$). For the four column
Shodex configuration intracolumn volume was ca four times as large
and the amount injected was increased proportionately. In addition,
for those samples where degradation appeared to occur, injection

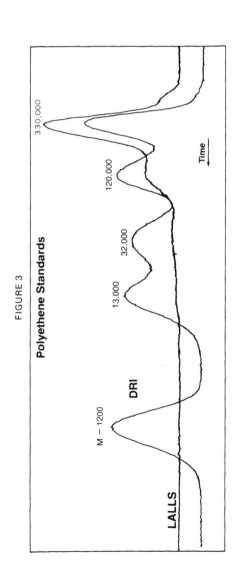

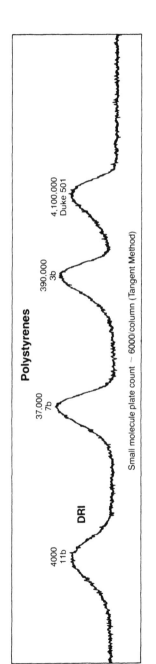

FIGURE 3. Typical blend calibration chromatograms.

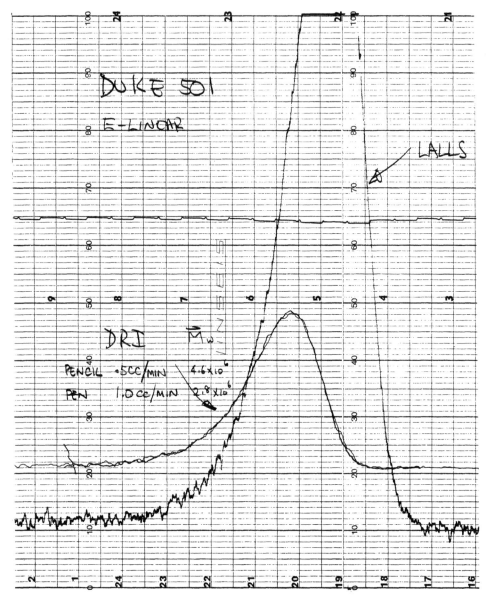

- Apparently for this column set, operating in the presence of degradation, dispersion effects predominate in determining peak shapes
- Better columns might show effects

FIGURE 4. Although flow rate changes change degradation, DRI profiles do not change significantly.

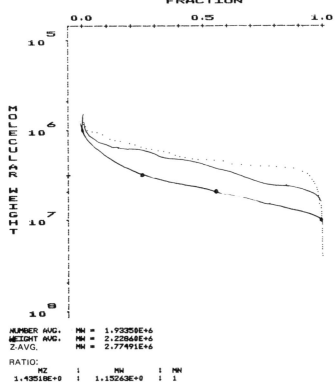

SAMPLE E-   6
DUKE 501 (10/31)(.5MG)(1.5ML)   [AIA077]
DATE: 2-11-79   TIME: 11:41:14
OPERATOR: JGR

V_E	Peak $\overline{Mw}$	Log $[\eta]M$				$\overline{Mw}$	$\overline{Mz}/\overline{Mw}$	$\overline{Mw}/\overline{Mn}$
5.01	2.41x10⁶	6.97	·········	1.5 ml/min	AIA 077	2.2x10⁶	1.43	1.15
4.87	3.05x10⁶	7.15	————	1.0 ml/min	AIA 090	2.8x10⁶	1.52	1.26
4.72	5.47x10⁶	7.58	——o——	.5 ml/min	AIA 094	4.6x10⁶	1.35	1.20

- Similarly, for actual numerical evaluation of MWD parameters there are no obvious changes

- These data are based on LALLS output

- If peak or overall $\overline{Mw}$ is used vs elution volume, both of which change with flow rate, the flow rate effects on log $[\eta]$ M vs elution volume are much diminished — see calibration curves — E-Linear Fig. 1 box. Degradation must occur early in this column set since polymer elutes ok for its real $\overline{Mw}$

- Unless one makes some kind of MW measurement on the eluant, nothing appears to be unusual, aside from "flow rate effects"

volume and concentration were varied over approximately a fourfold
range with the nominal values at mid range. Data are presented in
Table 1 and Figures 6 and 7.

Efforts were made to analyze for degradation in the absence of
columns. For the polystyrene sample Duke 501 no variation in
"plateau" Figure 1 or total integrated scattering intensity was observed in the flow rate range 0.2 to 1.0 ml/min. Thus it is believed
that at least for a polystyrene of 4.1×10^6 $\overline{M}_w$ the columns had to be
present to cause degradation. Results are difficult to obtain in the
absence of columns. Since solvent impurities, etc. elute at the same
time as the polymer, minor small molecule adsorption effects in the
DRI can cause bizarre traces. Absolute solvent refractive index
changes are not adequate to vary polymer dn/dc appreciably. Repeated injections had to be made to obtain reproducible DRI traces,
and even for the reproducible traces it cannot be confirmed that the
ratio of polymer to impurities is constant across the eluting sample.
LALLS traces were reproducible even when the DRI was not. If a large
enough sample is injected to give a plateau in concentration, however, this plateau ought to have the value of the original injected
sample. In fact the absolute values of concentration at the plateau,
calculated by proportioning the total DRI area to the total sample
weight injected are somewhat below the injected concentration. This
no doubt would vary with the nature of the impurities in the solvent.
Calculated $\overline{M}_w$ are high. Errors are of the order of 25%. Thus it is
the invariance of total integrated scattering which indicates little
or no degradation is taking place. Finally it is noted that the flow
perturbations caused in the KMX-6 cell (no columns present) were so
severe at 1 cc/min so as to preclude data collection at still higher
rates. With columns in place the KMX cell is protected from pump
stoke perturbations.

In general, noise and drift in the differential refractometer
was below ±2% and ca. ±4% in the light scattering signal. It is
possible to obtain large systematic deviations in $\overline{M}_w$ due to <u>excess</u>
scattering at high flow rates. The excess scattering appears to

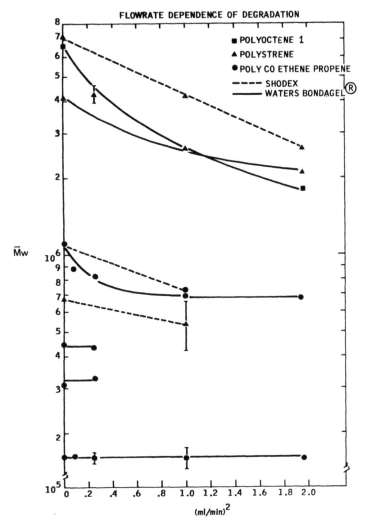

FIGURE 6. Flow rate dependence of degradation. Polymers of varied chemical structure and MWD are degraded at lowest practical flow rates--two column sets.

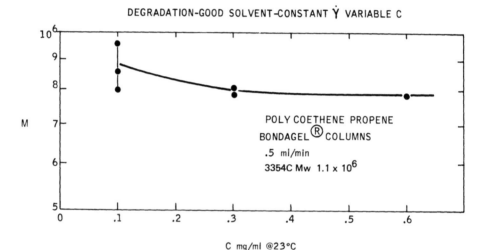

C mg/ml @23°C

Degradation increases with increasing polymer concentration for ethene — propene copolymers for which TCB is a good solvent.

It is uncertain whether degradation becomes negligible for this particular sample as C goes to zero.

For samples with Mz:Mw:Mn ~3:2:1 and $\bar{M}w$ ~1.5x10⁵ data become highly irreproducible above M ~7x10⁵ (ie; high Mw portions of the distribution) indicating even at very low concentration in the tails of the distribution degradation occurs.

We believe configurational effects involving knots may contribute to the concentration dependence.

FIGURE 7.   Concentration effects.

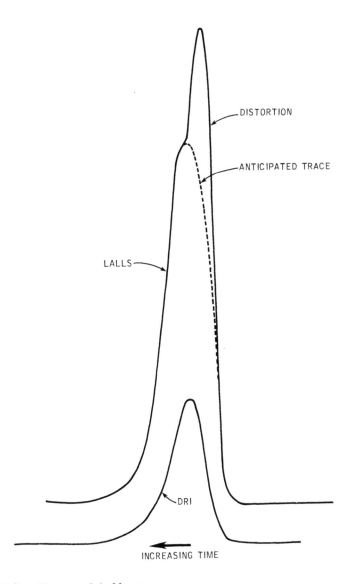

FIGURE 8.   Distorted Lalls trace.

arise from thermal inhomogenities due to improper temperature
control along the connecting tubing, which is not properly equili-
brated at the cell entrance at high flow rates. We use the heated
connecting line available from Chromatix, with extra insulation at
the cell and added heating tape at the GPC bulkhead. The effect
increases with the viscosity difference between solvent and elutant,
and may be exacerbated by scattering from the cell walls or windows
which does not appear to be in the field when using the viewing
microscope. $\overline{M}_w$ obtained at a series of injection volumes under such
conditions extrapolates to the statically obtained $\overline{M}_w$ at zero injection volume for samples which are known not to degrade under proper
operating conditions. The effect is more pronounced for narrow
distribution samples which pass the detectors at a higher concentration for a given injection amount. In its more extreme state the
LALLS trace becomes "bimodal" with excess scatter on the short elution time portion of the trace. No effect is seen in the DRI trace,
Figure 8. This problem may contribute to the apparent leveling of
$\overline{M}_w$ at high flow rates in Figure 6 [53].

A. Flow Characteristics and Potential for Degradation

The internal diameter of the connecting tubing used throughout was
0.009 in., 2.28×10^{-2} cm (228 μm). For flow rates of ca. 1 ml/min
the Reynolds number is 170 and flow is not turbulent at least in
the absence of polymer. The minimum shear rate, based on laminar
flow, is 1.4×10^4 sec^{-1}, and the Deborah number is ~ 20 τ_m, where
τ_m is the maximum relaxation time for the polymer. In the packed
columns a decision as to the correct correlating equation [25-28]
to use for pressure drop can be based on Reynolds number, i.e., the
group $D_p \, \rho \, v_0/\eta_0$ $(1 - \varepsilon)$ where D_p is particle diameter, ρ is density
and v_0 is the superficial velocity (see below), η_0 is the viscosity
and ε is the void fraction. Using $D_p = 1.0 \times 10^{-3}$ cm, $\rho v_0 \sim 0.5$
g/cm^2 sec, $\eta_0 = 5.4 \times 10^{-3}$ poise and $= 0.4$ one obtains ca. 2×10^{-1},
thus the flow in the packed columns is laminar. Internal diameter
of the columns is 4 mm and 8 mm for the E-Linear and Shodex,
respectively.

Pressure drop across the columns was ca. 1.0 bar. min/cm. ml
for the E-Linears and ca. one-fifth that for the Shodex. The major
portion of the pressure drop occurred in the columns rather than
ancillary connections and replacement of the $\sim$2-5 μm filters at the
column ends with clean elements prevents excess pressure drop due to
those elements alone.

If the column behaves as a bed of uniform diameter spheres in
laminar flow the pressure drop and column characteristics may be
approximated through Eq. 1, which is Darcey's Law [25]:

$$K = v_0 \eta L / \Delta P \tag{1}$$

where ΔP is pressure drop, K is the permeability of the medium, and
L the column length. A possible correlation for K is the Blake-
Kozeny-Carman [25] equation,

$$K = Dp^2 \varepsilon^3 / 180 (1 - \varepsilon)^2 \tag{2}$$

Data on polymer solutions [28] indicates significant perturba-
tion from η_0 is possible. We calculate lower bounds here but recog-
nize that the actual viscosity of the polymer solution, which
occupies only a small fraction of the packed bed until diluted an
order of magnitude by fractionation, may be significantly enhanced
due to the high elongational viscosity of the polymer.

From operation of the chromatograph we know that at a 0.5 ml/
min flow rate $v_0 \sim 0.2$ cm/sec with $\Delta P/L \sim 2.0 \times 10^5$ dynes/cm^2.
Therefore, K is a ca. 5.0×10^{-9} cm^2. If the particle diameter is
ca. 10 μm, ε is in the range of 0.55 to satisfy Eq. 2, a reasonable
result, although somewhat larger than anticipated. The ratio of the
particle diameter to superficial velocity should be a time on the
order of that needed to pass an expansion or contraction due to the
particles in the bed. This is on the order of 5×10^{-3} sec. If
molecular relaxation times, τ_m, are of this order or longer one can
expect substantial deformation [28,29-32] of the molecules in shearing and especially extensional flow regions in the columns. In
terms of Deborah number, De $\sim 200\tau_m$ in the packed bed. Elastic effects are noted at De ~ 0.1 [25,28].

There is substantial literature on τ_m which, in dilute solution scales [29] at $\tau_m \sim \eta_s R_F^3/kT$ or $\eta_s[\eta]M/kT$ where η_s is solvent viscosity, R_F is the radius of gyration, T is absolute temperature and k is Boltzman's constant. For our case $\eta_s/kT \sim 10^{11}$ sec/cm^3 so that R_F approaches 2×10^{-5} cm, τ_m is $\sim 10^{-3}$ sec. Experimental data [33] indicate similar values are reached by polystyrene at molecular weights of the order of 4×10^6. Polyolefins are easily within the 4000Å range as evidenced by the pore size needed for acceptable chromatography when molecular weight exceeds 10^6. Thus De exceeds 0.1 in typical GPC experiments on high molecular weight polymers.

In shearing flows an estimate of elastic effects can be obtained from the dilute solution Zimm theory, from which the stored energy per mole of molecules (which should be related to degradation [34]) is [25]

$$S = M^2 J_{eR} \eta_s^2 [\eta]^2 \dot{\gamma}^2 / RT \tag{3}$$

where M is molecular weight, J_{eR} is the reduced compliance, R the gas constant, and $\dot{\gamma}$ is the shear rate. For flexible molecules in poor solvents $J_{eR} \sim 0.2$. At 400 K, $S \sim 2 \times 10^{-16} M^2 [\eta]^2 \dot{\gamma}^2$ erg/molecule. For $M = 10^6$, $[\eta] = 10^3$ ml/g, $S \sim 2 \times 10^{-22} \dot{\gamma}^2$ erg/molecule. How large a stored energy is needed for bond breakage is yet to be determined [34,35]. The maximum in stored energy per bond peaks at the chain center but is still only of the order of the total stored energy divided by the number of backbone bonds. Even to attain 10^{-12} erg/molecule (50 kcal/mole) the shear rate would have to attain 10^5 sec^{-1}, an unlikely value. Thus, based on results of extrapolation of a linear theory to high strain rates, shearing flow in the columns does not appear to be the cause of the degradation. Extensional flow components will produce larger stresses [28] and stored energy at low strain rates, however. This is manifested by an apparent increase in $[\eta]$. Such flows have received study [35]. Comparison of estimated shear rates [51] with those observed to "cause" degradation in capillaries may be inappropriate because much of the capillary promoted degradation occurs in the entrance region and appears to be due to extensional flows.

B. Evidence for Extensional Flows

The following observations are pertinent to the enhancement of $[\eta]$
(i.e., $\eta_{solution}$).

For laminar flow, pressure drop through our piping should be
~ 2 bars, we obtain $\sim 3-4$ bars upstream of the KMX-6 cell. Pressure
increases caused by polymer injection in the absence of columns are
consistent with the shear viscosity of the solutions injected.

When columns are present two types of pressure traces are observed, those which exhibit "peaks" as the injected packet passes
certain portions of the columns and those which indicate a generalized rise across large distances in the system. Increases in pressure always are observed, for practical concentrations, when
degradation occurs (Fig. 9) [54].

First consider the general rise. For example, for an injection
of 3×10^{-5} g of sample 3354C (Table 1) at a flow rate of 1.5 ml/min
there is a smooth pressure rise lasting about 1 min amounting to an
increase of ca. 10% (e.g., 47 to 52 bars) from the pure solvent
pressure drop. This occurs ca. 6 mins after injection, for the
Shodex columns, when the polymer is significantly down the columns.
(Correlation of this time with column type and flow rate is under
study.) At the concentration injected viscosity increase caused by
the polymer must be linear in concentration, whatever the polymer
distribution in the columns. Thus $1.1 \sim \eta_{SOLN}/\eta_{TCB} = 1 + [\eta]Cf(\dot{\gamma})$,
where $f(\dot{\gamma})$ is an increasing function of $\dot{\gamma}$ and $[\eta] \cdot f(\gamma)$ may be regarded as an apparent intrinsic voscosity.

This must apply throughout the intersticies where dissipation
occurs and in which there is at least 10 cc of volume outside the
pores (Shodex). Thus $C_{average} \sim 3 \times 10^{-6}$ g/ml and $[\eta]_{Apparent}$ is
3×10^{4} ml/g. This is approximately 30 times the measured value in
shear. For such an apparent viscosity to be attained, strong extensional flow components must exist as this is the only plausible flow
which could be present in the columns and is also expected to cause
increased viscosity with increasing strain rate. In such flows
energy storage from flow startup is of the same magnitude as dissipation for flexible coils. For dumbbell models this can be shown by

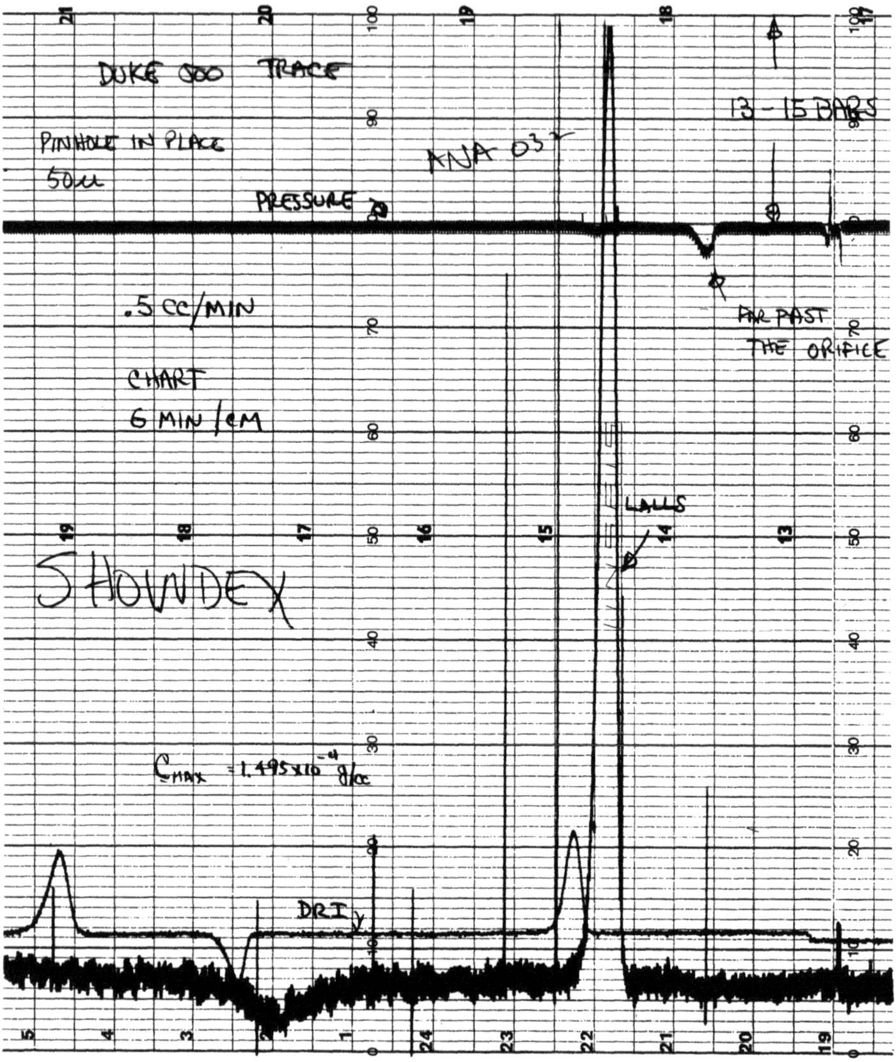

TRACE
PINHOLE IN PLACE
PRESSURE
13-15 BARS
.5 CC/MIN
CHART
6 MIN/CM
FAR PAST
THE ORIFICE
SHOWDEX

- For polymers which degrade, pressure peaks are apparent. These occur downstream of the first column entrance.
- They may occur at column exits, if this is so it might be possible to reduce the problem through appropriate packing size gradients, etc.
- However, for linear velocities of ~ .5 cm/sec passing 5μ, packing the flow through the constriction between particles takes only ~ 10^{-3} sec.

Maximum relaxation times of macromolecules reach this magnitude at M ~ 10^6 in solvents of TCB's viscosity (~ 5×10^{-3} Poise)
$\tau_m \sim \eta_s [\eta] M/_kT$

FIGURE 9. Pressure drop across columns is appropriate for solvent viscosity, packing size and ∿0.6 void fraction (seems large void fraction?)

integration of the appropriate expressions [25] for dissipation and coil deformation from t = 0 to t $\sim \tau_m$. Suppose the 5 bar ΔP (5 × 10^6 dynes/cm^2 or ergs/cc) may be considered to be stored by the stretched polymer. 3×10^{-5} g of 10^6 $\bar{M}_n$ polymer is 1.8×10^{13} molecules. 2.7×10^{-7} ergs/molecule is an enormous amount of stored energy when it is anticipated that carbon-carbon bonds have dissociation energy on the order of 6.8×10^{-12} ergs/bond (50 kcal/mole). It is adequate to supply the 50 kcal/mole activation energy each bond of a 10^6 MW polymer.

If instead of a general increase in pressure drop across the columns the pressure peak occurs due to passage of the polymer past a constriction, consider the following. In this case C in the previous analysis can be increased by a factor of $\sim$5-10. The apparent $[\eta]$ is reduced; but still exceeds $[\eta]_{shear}$ by 2-5 times, especially when one considers $[\eta]_{shear}$ would be a lower non-Newtonian value at these strain rates. Studies performed on a "low" molecular weight polymer indicate shear thinning dominates [17].

Thus, regardless of whether the pressure drop is localized or systematic, $[\eta]$ appears to be too large which indicates extensional flow.

It appears that any time perceptible pressure drops occur which cannot be ascribed to a gel fraction, it is likely that extensional flow components are present. These flows are effective in extending flexible molecules and thus in degrading them. Abundant energy is present to cause the degradation. Critical molecular weights and strain rates for coil extension may exist [29,30]. If these are exceeded the polymer may quickly degrade and pressure drop returns to normal.

C. Concentration Effects-Speculations

The data of Figure 7 indicate degradation is more severe at higher concentrations. The origin of this effect has not be conclusively demonstrated. Mechanistically it may be related to the effects of solvent thermodynamic quality.

Even at high dilution there appear to be large effects of solvent quality on degradation [36-44,50]. Thus, polymers degrade more in poor solvents. We believe the reason for this may be related to the concentration of knots (Fig. 10) in these two solvent regimes [45-47]. Knot concentration increases as $M^{1/2}$ and ln M in poor and good solvents, respectively. Knots, when drawn tight (e.g., $\sim$10 main chain carbons) should facilitate breakage by stress raising. A tenuous analogy can be made with the effects of knots on rope strength. We plan further efforts in this area. Similar effects on chain dimensions and knot concentration should be produced, in good solvents, by increasing polymer concentration. Thus as chain dimensions contract upon polymer concentration increase, knot concentration should increase.

In any event the practical result appears to be that experiments should be run at as low a concentration as possible, however at the lowest practical concentration degradation persists.

From the previously cited studies it is suggested that good solvents (of low viscosity) would also minimize degradation.

Although literature data are somewhat contradictory generally polymers degrade more in poor solvents than in good ones and the concentration dependence is much diminished in poor solvents. Some factor other than shear stress or stored energy/ molecule must be important. Chemical effects due to the solvent could be important but it is unlikely that poor solvents would always be chemically best for degradation.

Knot concentration/polymer coil varies as:
$M^{1/2}$ poor solvents.
In M good solvents.

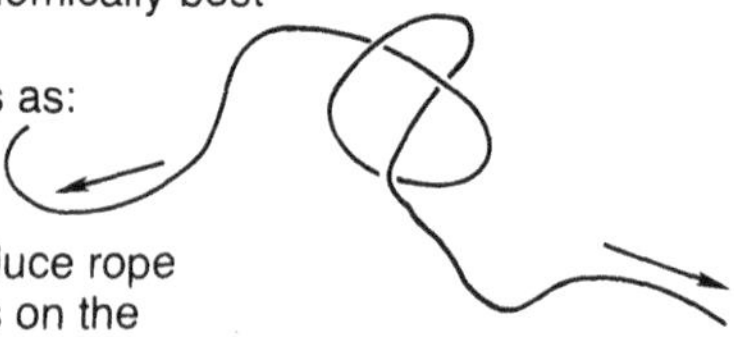

By analogy with the fact that knots reduce rope strength we speculate the effect exists on the molecular level. Tensile stresses in the chain could be more effective in distorting particular bonds if knots were present. Efforts are underway to get appropriate calculations performed.

(Thanks to E. Kresge for reminding us to look in knot file.)

FIGURE 10. Origins of concentration effects.

D. Orifice Flows

A limited amount of data has been collected wherein orifices have been installed in the piping upstream of the columns. The flow of polymer solutions in orifices is being studied [35]. For conventional 228 μm tubing pinholes in the range 25 to 5 μm radius should provide appropriate kinematics to produce degradation. Such pinholes are commercially available (Energy Technology Inc., San Luis Obispo, Calif.)

        Preliminary experiments using a 25 μm radius pinhole (4.6:1 contraction) produced no degradation beyond that already occurring in the Shodex columns which were the ones in place during the experiment. No pressure peak that could be attributed to the orifice was observable (Fig. 9). The calculated inertia dominated ΔP for the solvent is less than a bar for this type contraction. This gives some indication of the mismatch that can be tolerated in tubing unions, etc. without degradation occurring. Efforts to collect data using smaller pinholes have been thwarted due to leaks, plugging, and a lack of time. Further efforts are planned.

III. DISCUSSION

The data of Table 1 and Figures 5-7 show convincingly that degradation occurs in routine GPC separations on microparticulate substrates. Without on line molecular weight detection it is difficult, rigorously, to demonstrate that degradation has occurred. Since some degradation occurs down to the lowest practical flow rates (ca. 0.2 ml/min) it appears that there is a limit to the rate at which GPC data can be collected on high molecular weight polymers. This rate must be on the order of a hundred minutes. Columns must be operated such that (for solvents of similar thermodynamic quality) $Dp/v_0 > \eta_0 R_F^3/kT \approx b[\eta]M$, where b is a constant for a given solvent and temperature. That such a correlation could exist might explain some of the success of MWD measurements based on hydrodynamic volume, even though degradation is occurring. It can only be a first approximation due to chemical and long chain branching effects.

The trends in high speed chromatography have been to lower Dp so that columns could be shortened while maintaining ν_0, resolution and efficiency. For typical commercial polymers of interest (e.g., polyolefins) in good solvents a limit appears to have been exceeded. Thus with $\nu_0 \sim 0.2$, Dp ~ 10 μm for polyethylene, degradation ensues at $< [\eta] \sim 10^3$ ml/g, M $\gtrsim 10^6$ or b $= 2 \times 10^{-7}$ sec/cm^3. Correlation of the rheological properties of such polymers with GPC results might be impossible since the high molecular weight portions of MWDs are made to appear equivalent by degradation.

Having concluded that degradation does occur one can enquire what variables other than particle size and flow rate should be modi-fied to obtain optimum separation and minimal degradation.

A compromise must be reached in temperature. Raising tempera-ture lowers η_s and shortens τ_m, but eventually thermal mechanisms of degradation will intercede.

Increasing polymer concentration not only enhances viscosity but will cause a decrease in chain dimensions with a concomitant increase in knot concentration. The experimental result of Figure 7 indicates low concentration is best regardless of the explanation. Good solvents of low viscosity should be used.

IV. CONCLUSIONS

The importance of mechanochemical degradation in the GPC process is demonstrated. There is a limit for the speed at which MWD data can be accumulated and it is on the order of 10^2 min for current column technology.

Efforts to improve efficiency and resolution in GPC by reducing packing particle size (to 5-15 μm) have produced flow kinematics which degrade polymer of molecular weight in the range of practical significance, e.g., $[\eta]$M $\sim 10^7$ dl/mole or M $\sim 10^6$ for polyethene (the literature [51] now indicates problems with 37-75 μm particles).

Since MWD and long chain branching are expected to change the way polymers degrade, it will be fortuitous if high speed GPC can be used to characterize broad MWD plastics and elastomers even on a relative basis.

Researchers are obligated to show degradation does not take place in GPC results they present.

Efforts should be made to determine whether high pressure drops are confined to certain portions of the GPC column; and if so, to alter packing in such regions (e.g., exits).

Efforts should be made to define the appropriate particle size and shape to optimize efficiency and resolution while minimizing degradation, e.g., avoiding extensional flows.

Even with optimized-packing, good solvents of low viscosity should be used as mobile phases.

REFERENCES

1. A. Ouano, J. Chromatog., 118, 303 (1976).

2. A. Hamielec, A. Ouano, L. Nebenzahl, J. Liq. Chromatog., 1, 527 (1978).

3. A. Hamielic and A. Ouano, J. Liq. Chromatog., 1, 111 (1978).

4. T. MacRury and M. McConnell, J. Appl. Pol. Sci., 24, 651 (1979).

5. D. Aaxelson and W. Knapp, J. Appl. Pol. Sci., 25, 119 (1979).

6. T. Kato, A. Kanda, A. Takahashi, I. Noda, S. Maki, and M. Nagasawa, Polymer J., 7, 575 (1979).

7. R. C. Jordan, J. Liq. Chromatog., 3, 439 (1980).

8. A. Cooper, J. Johnson and A. Bruzzone, Eur. Pol. J., 9, 1381 (1973).

9. S. Mori, J. Appl. Pol. Sci., 21 1921 (1977).

10. J. Knox and F. McLennon, J. Chromatog., 185, 289 (1979).

11. J. Dawkins, Pure Appl. Chem., 51, 1473 (1979), and reference therein.

12. H. Mahabadi and A. Rudin, Polymer J., 11, 123 (1979).

13. J. Janca, Anal. Chem., 51, 637 (1979).

14. J. Lecourtier, R. Audebert, and C. Quivoron, Pure Appl. Chem., 51, 1483 (1979).

15. J. V. Dawkins and G. Yeadon, J. Chromatog., 188, 333 (1980).

16. A. R. Cooper, J. Liq. Chromatog., 3, 393 (1980).

17. J. Janca, J. Chromatog., 187, 21 (1980).

18. E. Slogowski, L. Fetters, and D. McIntyre, Macromolecules, 7, 394 (1974).

19. H. Wagner and C. Hoeve, J. Pol. Sci., A2, 9, 1763 (1971).

20. W. Park and W. Graessley, J. Pol. Sci. Physics, 15, 71 (1977) for an example of tailing in series detectors.

21. H. Wagner and P. Verdier, J. Res. Nat. Bureau. Stds., 83, 195 (1978).

22. H. Wagner and C. Hoeve, J. Pol. Sci. Phys., 11, 1189 (1973).

23. A. Barlow, L. Wild, and R. Ranganath, J. Appl. Pol. Sci., 21, 3319 (1977).

24. H. Coll and D. Gilding, J. Pol. Sci., A2, 8, 89 (1970).

25. R. Bird, R. Armstrong, and O. Hassager, Dynamics of Polymeric Liquids, Wiley, New York, 1977.

26. Z. Kemblowski and M. Michniewicz, Rheol. Acta, 18, 730 (1979).

27. I. F. Macdonald, M. El-Sayed, M. Kow, and F. Dullien, Ind. Eng. Chem. Fund., 18, 199 (1979).

28. C. Elata, J. Burger, J. Michlin, and V. Taskerman, J. Phys. Fluids, 20, S49 (1977).

29. P. G. deGennes, Scaling Concepts in Polymer Physics, Cornell U. Press, Ithaca, New York, 1979, p. 186ff.

30. A. Peterlin, Die Makromolekulare Chemie, 453 (1975).

31. J. Ferry, Viscoelastic Properties of Polymers, Wiley, New York, 1970.

32. J. Janeschitz-Kriegl, Adv. Pol. Sci., 6, 170 (1969).

33. G. Berry, J. Chem. Phys., 44, 4550 (1966).

34. G. Ver Strate, C. Cozewith, and W. Graessley, Polymer Preprints, 20, 149 (1979).

35. R. Armstrong, S. Gupta, and O. Basaran, Pol. Eng. Sci., 20, 466 (1980).

36. A. Casale and R. Porter, Polymer Stress Reactions, Academic Press, New York, 1979.

37. J. Culter et al., J. Appl. Pol. Sci., 16, 3381 (1972).

38. J. Knight, Mechanical Shear Degradation of Polymers: A Review, 1976 NTIS Publication AD A038 139 and references therein.

39. A. Nakano and Y. Minoura, Macromolecules, 8, 677 (1975).

40. P. Sheth, J. Johnson, and R. Porter, Polymer, 18, 741 (1977).

41. D. Hunston and J. Zakin, Polymer Preprints, 19, 430 (1978).

42. C. Chen and E. Sheppard, Polymer Preprints, 19, 424 (1978).

43. P. Leopairat and E. Merrill, Polymer Preprints, 19, 418 (1978).

44. J. F. S. Yu, J. L. Zakin, and G. K. Patterson, J. Appl. Pol. Sci., 23, 2493 (1979).

45. A. Vologodskii et al., Sov. Phys. JETP, 39, 6 (1974).

46. H. Frisch and E. Wasserman, JACS, 83, 3789 (1961).

47. F. Brochard and P. deGennes, Macromolecules, 10, 1157 (1977).

48. Y. Yu, J. Zakin, and G. Patterson, J. Appl. Pol. Sci., 23, 2493 (1979).

49. R. Ting and D. Hunston, J. Appl. Pol. Sci., 21, 1825 (1977).

50. J. Zakin and D. L. Hunston, J. Appl. Pol. Sci., 22, 1763 (1978).

51. C. Huber and K. Lederer, Polymer Lett., 18, 535 (1980). This paper, which appeared after submission of our manuscript to Polymer Preprints, 21, 196 (1980), independently concludes mechanical degradation is significant in a polyisobutylene sample using on line light scattering.

52. R. Bressau, Chromatogr. Sci., 13, 73 (1980).

53. It was suggested at the symposium that this may also be a lens effect caused by concentration as well as temperature gradients in the cell. Whatever the cause the main beam is deflected so that light is collected at lower effective angle and intensity increases. The same effect, if it were decreasing intensity, and causing apparent $\overline{M}w$ decrease should have existed in the experiments with no columns but no $\overline{M}w$ decrease was observed.

54. R. Cressely and R. Hocquart, Polymer Preprints, 22, 120 (1981). This paper presents interesting results on the effects of a series of convergent-divergent flows on polymer coil expansion.

GEL PERMEATION CHROMATOGRAPHIC ANALYSIS OF POLY(2-METHYLPENTENE-1
SULFONE) AND POLY(BUTENE-1 SULFONE): THE INFLUENCE OF POLYMER-
COLUMN AND POLYMER-SOLVENT INTERACTIONS ON ELUTION BEHAVIOR

Gary N. Taylor
Molly Y. Hellman
Larry E. Stillwagon

Bell Laboratories
Murray Hill, New Jersey

I. INTRODUCTION

Aliphatic polysulfones are a class of polymeric materials having an
alternating structure in the most general sense consisting of polar
SO_2 groups and nonpolar aliphatic groups R. The general structure
is given in Structure 1.

$$- R - SO_2 - R - SO_2 - R - SO_2 - \qquad (1)$$

These polymers are prepared by free radical polymerization [1] using
either gamma ray, UV, or thermal initiation. Because of their
relatively weak backbone bonds, aliphatic polysulfones are readily
degraded by high energy radiation. This property permits them to
function as sensitive resists when exposed to ionizing radiation
such as that encountered in electron lithography. Two specific
materials employed as electron resists are poly(butene-1 sulfone)
(Str. 2) and poly(2-methylpentene-1 sulfone) (Str. 3).

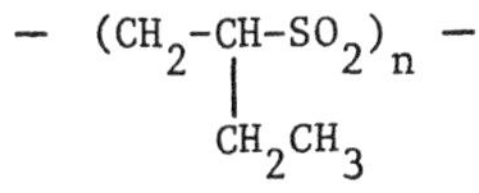

$$- (CH_2-CH-SO_2)_n - \qquad\qquad - (CH_2-\overset{\overset{\displaystyle CH_3}{|}}{\underset{\underset{\displaystyle CH_2CH_2CH_3}{|}}{C}}-SO_2)_n -$$

$$\underset{\underset{\displaystyle CH_2CH_3}{|}}{}$$

(2) 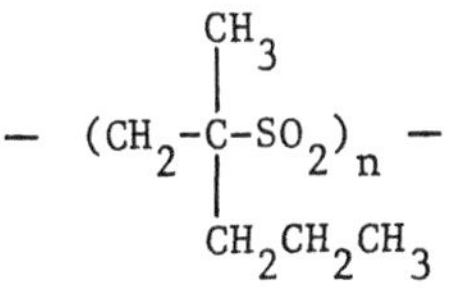   (3)

237

For the former which is often called PBS, high energy radiation
causes chain cleavage and a reduction in molecular weight. This
and other factors cause the rate of dissolution to be faster in the
irradiated regions [2]. An accurate knowledge of molecular weight
and molecular weight distribution is essential to obtaining reproducible lithographic properties.

Sulfone (Str. 3), often called PMPS, finds use as a resist in
combination with novolac resins [3]. It functions as a dissolution
inhibitor. Upon irradiation it undergoes depolymerization thus con-
verting it to gaseous SO_2 and 2-methyl-1-pentene products. Its
properties in composite samples are dependent on molecular weight
and molecular weight distribution. Again an accurate knowledge of
these parameters is required to optimize lithographic properties.

The measurement of $\overline{M}_w$ and $P = \overline{M}_w/\overline{M}_n$ by gel permeation chromatography for both PBS and PMPS has had mixed results. Bowden and
Thompson [4] reported good success with GPC analyses for PBS in THF
solvent. We have had variable results with PBS. Elution was always
accomplished, but the elution volumes were sometimes variable. Ouano
and coworkers [5,6] reported unusual results for another poly(sulfone). This was poly([2.2.1]bicyclohept-2-ene sulfone). They
studied the molecular properties of this sulfone by GPC and LALLS,
low angle laser light scattering. They found that $\overline{M}_w$ determined by
GPC using the universal calibration method was 6-7 times greater
than that measured by light scattering. The solvent was $CHCl_3$.
Similar results were obtained using cyclohexanone. Originally Ouano
attributed this to polymer-solvent association [5], but later ruled
this out on the basis of the Mark-Houwink coefficients. In addition
the authors found considerable "forward skew" in the chromatograms.
This is usually associated with nonequilibrium chromatography. However, the cause of such effects was not delineated. Hindered diffusion of the polymer into the gel pores due to a rodlike polymer
structure is one possibility.

For PMPS the chromatographic situation is more complicated. On
one set of μ-Styragel columns using THF as the solvent, samples with

a variety of molecular weights were not eluted. On an identical
machine with μ-Styragel columns and THF solvent, the samples were
eluted but with variable results. The present account describes
the results of experiments that were designed to elucidate the
causes of these anomalous GPC results encountered with polysulfones.

II. EXPERIMENTAL

All GPC analyses were conducted at room temperature on Waters Model
244 high pressure liquid chromatographs. Two column types were
used: (1) a series of five μ-Styragel (Waters) columns having pore
sizes of 10^6, 10^5, 10^4, 10^3, and 5×10^2 Å, and (2) two Zorbax
(Dupont) silica columns functionalized with trimethylsilyl groups
and having pore sizes of 60 Å and 1000 Å. The eluent peaks were detected with a differential refractive index detector. For most runs
an internal standard, benzene [7], was used to correct for small
variations in flow rate. The main solvent used was tetrahydrofuran
obtained from Waters. It was a specially purified grade having no
UV absorbing contaminants. Ethyl acetate, toluene, chloroform, and
carbon tetrachloride were Fisher analytical grade reagents and were
used as received. H_3PO_4 was Baker reagent grade. Oxalic acid,
acetic acid, maleic acid, p-toluenesulfonic acid, and sulfuruc acid
were Fisher analytical grade reagents. Chloroacetic acid and picric
acid were obtained from Aldrich. Samples of PMPS and PBS were obtained from E. M. Doerries and Mead Chemical Co.

III. RESULTS AND DISCUSSION

Initially, PMPS in THF solution was analyzed on three sets of μ-
Styragel columns. On set A a continuously-stirred solvent reservoir
was used only with THF solvent. On sets B and C an unstirred solvent
reservoir was used. Column set B had a previous history which included the use of at least six different solvents including dimethyl
formamide. Column set C had no previous use. PMPS solutions in THF
injected on column sets B and C did not exhibit elution of any

material. On column set A PMPS initially was eluted satisfactorily. However, the elution curve shown in Figure 1a could not be satisfactorily reproduced. It was surmised that water adsorbed from the atmosphere was responsible for these affects. To test this, 1% water was added to the THF mobile phase. This resulted in elution which produced a broad molecular weight distribution and elution at lower V_e values corresponding to higher molecular weight as shown in Figure 1b. Upon changing back to pure THF solvent the elution curve shown in Figure 1c was obtained. This curve was intermediate between those of 1a and 1b and slowly changed with continued running until curve 1a was obtained.

On the basis of these results we speculated that water was absorbed on the column thereby preferentially solvating active sites that were preferred by the PMPS in the absence of water. One possible group on the polymer which could be responsible for preferential adsorption at polar sites is the sulfinic acid end group which may be present at chain ends or branch points on the polymer. If this were true then sufficiently strong acids would cause protonation of the "basic" sites on the column thought to chemisorb the PMPS.

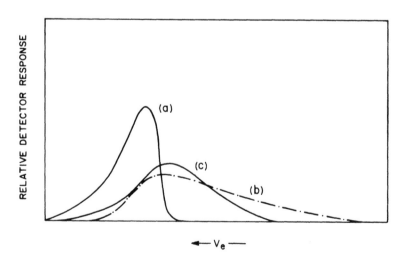

FIGURE 1. Gel permeation chromatograms of PMPS in (a) THF, initial run; (b) THF with 1% water added; (c) THF rerun after flushing with THF.

This would shift the equilibrium outlined in Scheme 1 to favor polymer in solution where P $\rightarrow$ SO$_2$H is the polymer containing sulfinic
acid groups, B is a basic site on the column, and HA is an acid of
ionization constant K. To test this a series of GPC runs were conducted using column sets B and C, THF solvent, and additional known
amounts of acids varying in both concentration and ionization constant. The results are summarized in Table 1. Note the absence of
elution for the very weak acids and elution for those acids having
a pK$_a$ < 2.12. Also note that acid concentration has a considerable
effect on V$_e$ and $\overline{M}_w$. Thus, the elution volume decreases and the
molecular weight increases as the concentration of H$_2$SO$_4$ in THF
is increased. GPC traces are shown for three concentrations of
H$_2$SO$_4$ in Figure 2. The influence of acid strengths at a fixed acid
concentration is presented in Figure 3.

Taken separately these results may indicate the validity of the
basic site scheme outlined above. However, there are several flaws
in this logic. Such a scheme would require an orderly rather than a
random dependence of V$_e$ on pK$_a$. Also, the polydispersity data
varies widely. Perhaps the best indication of other factors is the

TABLE 1

Effect of Acid Strength and Concentration
on the GPC Analysis of PMPS Sample
23DE in THF

Acid	pK$_a$	Conc. (N)	V$_e$(ml)	$\overline{M}_w$(g/mole)	P
CH$_3$CO$_2$H	4.75	0.20	No PMPS	Elution	--
ClCH$_2$CO$_2$H	2.85	0.091	No PMPS	Elution	--
H$_3$PO$_4$	2.12	0.108	177.89	124,000	1.59
Maleic Acid	1.85	0.091	178.77	120,000	1.35
p-Toluenesulfonic Acid	0.90	0.091	161.95	569,500	2.13
Picric Acid	0.38	0.091	162.58	508,000	2.53
H$_2$SO$_4$	<0	0.094	179.50	178,700	1.45
	<0	0.187	165.80	408,500	1.69
	<0	1$^+$%	161.30	663,700	1.32

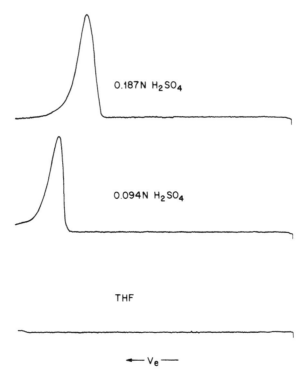

0.187N H₂SO₄

0.094N H₂SO₄

THF

◄— Vₑ —

FIGURE 2. Gel permeation chromatograms of PMPS in THF to which
H_2SO_4 was added.

failure of the three strong acids, p-toluenesulfonic, picric and
sulfuric acid, to give increasing $\overline{M}_w$ values at equal proton concentrations corresponding to 0.091, 0.019, and 0.187 N, respectively.
Each acid thus exhibits its "own" solvent effect.

Even more distrubing is the fact that upon changing back to
pure THF solvent, no PMPS elution was observed. This indicates that
there are not any "basic" sites on the column which cause chemisorption of PMPS because once converted to the acid form these would not
be expected to revert to the basic form unless base were passed
through the column. Thus, it appears that the acids modify the solvent and elution is the result of this solvent effect alone.

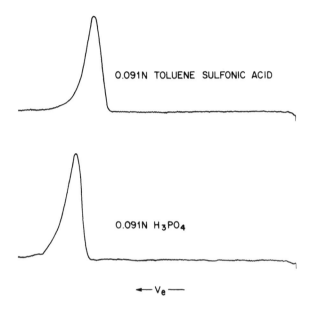

FIGURE 3. Influence of acid strength on elution behavior of PMPS in THF.

Considering that such "basic" sites are not involved, we were curious as to whether the carrier solvent itself was responsible. THF is known to complex strongly to cations and may influence the equilibria in Scheme 1 favoring chemisorption.

$$P - SO_2H + B \;\underset{<}{\overset{>}{=\!=}}\; P - SO_2^- + BH^+ \tag{4}$$

(solution)   (chemisorbed)

$$H_2O + HA \;\overset{K}{\underset{\Longleftarrow}{=\!\!=\!\!=}}\; H_3O^+ + A^- \tag{5}$$

$$P - SO_2^- + H_3O^+ \;\underset{<}{\overset{>}{=\!=}}\; P - SO_2H + H_2O \tag{6}$$

Scheme 1

Nonpolar solvents on the other hand should favor the unionized form in Structure 4. Consequently, we tried nonpolar solvents which still satisfactorily swell the crosslinked poly(styrene) beads (μ-Styragel)

of the support. The results are given in Table 2. With the
exception of chloroform all solvents failed to elute PMPS. Typical
elution curves are presented in Figure 4. Chloroform is exceptional
because it contains ethanol which causes elution of PMPS. Addition
of ethanol to other solvents likewise permits elution. Typical results are presented in Table 3. The effect of ethanol in THF is
considerably attenuated relative to the effect of water in the same
solvent. Once again the results point to a solvent effect as being
responsible for the widely varying elution behavior. For example,
note the extraordinary variation in peak broadening as indicated by
the polydispersity values in Table 3.

In summary, the previous results indicate that retention of
PMPS on the columns is not caused by interaction at specific basic
sites. Successful elution in the presence of strongly acidic or
hydrogen bonding compounds added to moderately polar and nonpolar
solvents suggests that PMPS is absorbed or adsorbed by the μ-Styragel
in their absence. Such sorption would be removed by phases incapable
of sorbing PMPS. Consequently, we tested Zorbax columns obtained
from Dupont. These columns contain silica spheres functionalized
with trimethylsilyl groups to remove surface hydrogen bonding sites.
Using these columns we obtained elution in ethyl acetate and
tetrahydrofuran. Typical chromatograms obtained in ethyl acetate
solvent for samples 18DE and 23DE are shown in Figure 5.

TABLE 2

Solvent Effects on the
GPC Analysis of PMPS Sample 23DE

Solvent	ε	V_e (ml)	$\overline{M}_w$ (g/mole)	P
Tetrahydrofuran	7.58	No PMPS	Elution	--
Ethyl Acetate	6.02	No PMPS	Elution	--
Chloroform	4.80	157.12	687,300	7.79
Toluene	2.38	No PMPS	Elution	--
Carbon Tetrachloride	2.24	No PMPS	Elution	--

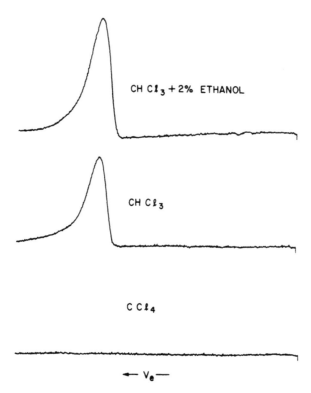

FIGURE 4. Gel permeation chromatograms of PMPS in nonpolar solvents
and the effect of added ethanol on elution behavior.

TABLE 3

Effect of Ethanol on the GPC
Analysis of PMPS Sample 23DE

Solvent	Ethanol conc. (%)	V_e (ml)	$\overline{M}_w$ (g/mole)	P
Tetrahydrofuran	0	No PMPS	Elution	--
Tetrahydrofuran	5	191.50	41,700	1.40
Tetrahydrofuran	10	186.70	61,424	1.43
Chloroform	1	157.12	687,300	7.79
Chloroform	2	158.86	650,300	3.72

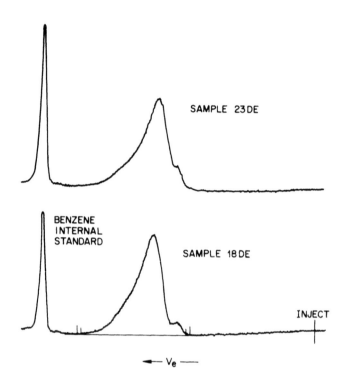

FIGURE 5. Gel permeation chromatograms of PMPS in ethyl acetate for two samples on Zorbax columns. On μ-Styragel columns no elution was observed.

A series of five samples were analyzed in this solvent. The samples were prepared under a variety of conditions. The analytical results are summarized in Table 4. The columns have sufficiently high resolution to resolve what appears to be bimodality in four of the five samples. A unimodal distribution was found only for the lowest $\overline{M}_w$ material. Resolution is improved in THF solvent as indicated by the chromatogram shown in Figure 6. A log-log plot of [η] versus $\overline{M}_w$ (Fig. 7) is not linear for $\overline{M}_w$ values greater than 1×10^6 g/mole indicating that the pore sizes are not sufficiently large to resolve the very high molecular weight species.

The bimodal nature found for four of the five PMPS samples in Table 4 was surprising to us. PMPS is prepared by a free radical

TABLE 4

GPC Analysis of PMPS on Zorbax
Columns in Ethyl Acetate

Sample #	$[\eta]$ (ml/g)	$\overline{M}_w$ ($\times 10^{-6}$ g/mole)	P	Modality
2DC	410	1.43	2.57	Bi
42DD	140	0.16	1.81	Uni
18DE	180	0.95	3.38	Bi
23DE	265	1.31	4.33	Bi
31DE	400	1.38	2.73	Bi

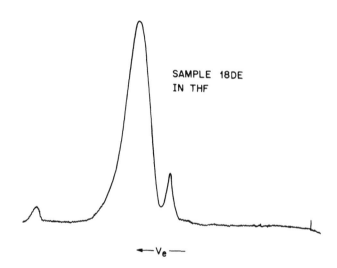

SAMPLE 18DE
IN THF

$\longleftarrow V_e \longrightarrow$

FIGURE 6. Gel permeation chromatogram of PMPS sample 18DE in THF on Zorbax columns.

technique which should not afford bimodal material. If the distributions for the bimodal samples were in fact unimodal what could possibly cause the bimodality? One possibility is that the silica Zorbax columns may not be completely silylated. The remaining OH groups on the support could hydrogen bond to the PMPS thus causing it to elute later. The material in the high molecular weight tail

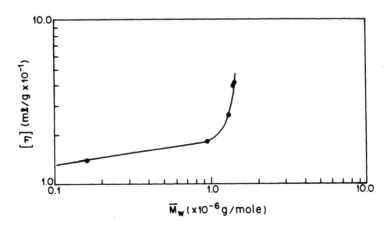

FIGURE 7. Log-log plot of [η] in ethyl acetate versus poly(styrene) equivalent molecular weight in ethyl acetate determined on Zorbax columns.

would not undergo such interactions because its very high molecular weight excludes it from most if not all of the pores. Continued silylation should eliminate the bimodality if polymer–column interactions were responsible.

In Figure 8 we show chromatograms obtained in ethyl acetate for sample 18DE of PMPS. The initial untreated sample indicated by a zero in the figure is bimodal. Each successive sample was analyzed after injecting 100 μl of a 3:1 mixture comprised of pyridine and bistrimethylsilyl acetamide, a well known silylating agent. Note that the shoulder is not present after the first treatment. Successive treatments shift the curves to higher values of $\overline{M}_w$ (lower V_e) but do not change the elution behavior of poly(styrene) standards. Upon completion of treatment 4 and continued pumping of solvent through the system, the PMPS chromatograms slowly revert to the original bimodal chromatogram. These results can be ascribed to polymer–support interactions at hydrophilic sites on the support. Thus polymer–support interactions are extraordinarily important in determining the elution behavior of PMPS even on columns for which sorption is nominally absent.

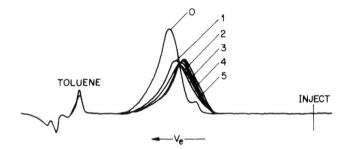

FIGURE 8. Gel permeation chromatograms of PMPS sample 18DE in ethyl acetate on Zorbax columns before treatment 0 with *bis*-trimethylsilyl acetamide-pyridine and after successive treatments 1-5.

Silylation does not provide a means for permanently modifying the support surface of the Zorbax columns. We believe that this arises because of the presence of especially labile sites which upon silylation are readily hydrolyzed. Are there any other methods which might safely protect the polymer from these sites? One general method well known to those involved in this art is to incorporate additives in the elution solvent containing two or more hydroxyl functional groups. Among these are diols, polymeric diols and triols. We have evaluated propylene glycol in this study. Our results are presented in Figure 9 for PMPS in ethyl acetate and in Tables 5 and

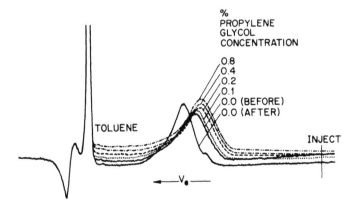

FIGURE 9. Gel permeation chromatograms in ethyl acetate of PMPS sample 18DE on Zorbax columns and in ethyl acetate containing various percentages of propylene glycol.

TABLE 5

Influence of Added Propylene Glycol
on the Elution of PMPS Samples 18DE
and 46DE in Ethyl Acetate

Sample	%PG in EA	M_w (g/mole)	P
18DE	0.0	4,160,000	8.5
	0.1	4,720,000	10.2
	0.2	7,180,000	12.0
	0.4	8,470,000	10.1
	0.8	10,050,000	10.5
	0.0	1,840,000	4.8
46DE	0.0	7,390,000	6.05
	0.1	8,650,000	4.94
	0.2	12,900,000	5.16
	0.4	14,200,000	5.99
	0.8	15,700,000	4.72
	0.0	2,630,000	3.28

TABLE 6

Influence of Added Propylene Glycol
on Elution of PBS Samples MP20 and
MP20-1MRAD in Ethyl Acetate

Sample	%PG in EA	$\overline{M}_w$ (g/mole)	P
MP20	0.0	1,640,000	4.21
	0.1	2,080,000	4.33
	0.2	2,650,000	3.83
	0.4	3,080,000	4.10
	0.8	3,680,000	4.74
	0.0	1,260,000	2.99
MP20-	0.0	261,000	3.23
1MRAD	0.1	373,000	3.20
	0.2	554,000	5.26
	0.4	665,000	5.23
	0.8	851,000	7.23
	0.0	371,000	3.72

6 for PMPS and PBS samples, respectively. In the figure we see that
addition of even 0.1% of propylene glycol has eliminated the high
molecular weight peak seen originally (Fig. 7). It is also absent
in the trace obtained before treatment which was run immediately
after the fifth silylation in Figure 8. Note that the elution
volume V_e decreases and calculated $\overline{M}_w$ increases as the diol concentration increases. This occurs for both PMPS and PBS and for two
different molecular weight samples of each material as shown in
Tables 5 and 6. The $\overline{M}_w$ and P values are outrageously high especially for the 0.8% concentration. It appears that although we have
eliminated the polymer-column interaction by this method, we have
introduced an even greater and more detrimental affect by introduc-
ing a hydrogen bonding dopant, propylene glycol. We believe that
aggregation might be caused by such interactions and are currently
checking this possibility by light scattering measurements.

Is there any way of permanently protecting the labile sites on
the Zorbax columns? One way which might work is to react these
sites with a silane material which is very bulky. The reverse hydrolysis step would then be hindered considerably and prevent reaction
at ambient conditions. Thus we decided to react the Zorbax columns
with pyridine solutions of esters having Structure 7.

$$((CH_3)_3SiO)_3SiOCR \overset{O}{\overset{\|}{}} \tag{7}$$

The results are summarized in Figures 10 and 11 and Tables 7 and 8.
The chromatograms of PMPS and PBS, respectively, were run after injecting four 200 μl samples of Structure 4 in pyridine and running
them through the Zorbax columns maintained at 50°C using ethyl acetate solvent carrier. Temperature was maintained in the Waters 150-C
machine. The columns were cooled to room temperature and the ethyl
acetate was passed through the columns for 16 hours. Another sample
of each polymer was analyzed (sample 2). Hydrolysis in the 1% H_2O
solution was again conducted for 48 additional hours and the polymers
were analyzed once again. In Figure 10 the elution curves for all

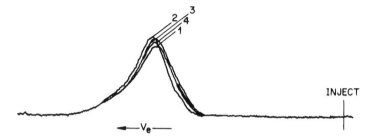

FIGURE 10. PMPS chromatograms after treatment with hindered silane
(1) and after hydrolytic treatment for 16 hrs (2) and 64 hrs (3) (4).

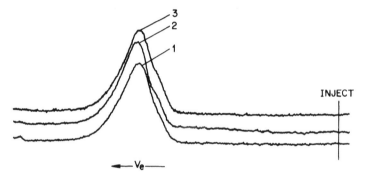

FIGURE 11. PBS chromatograms after treatment as in Figure 10 (1) and
after hydrolytic treatment for 16 hours (2) and 64 hours (3).

TABLE 7

Effect of Silylation Treatment
and Hydrolysis Conditions on the GPC
Analysis of PMPS Sample 18DE
in Ethyl Acetate

Sample #	Hydrolysis Hydrolysis Time (hrs)	$\overline{M}_w \times 10^{-6}$ (g/mole)	P
1	0	3.89	7.87
2	16	2.74	5.74
3	64	3.56	7.01
4	64	3.61	7.70

TABLE 8

Effect of Silylation Treatment
and Hydrolysis Conditions on the GPC
Analysis of PBS Sample MP20
in Ethyl Acetate

Sample #	Hydrolysis Time (hrs)	$\overline{M}_w \times 10^{-6}$ (g/mole)	P
1	0	1.46	3.25
2	16	1.27	2.96
3	64	1.50	3.18

three PMPS samples are nearly identical. No high molecular weight peak was observed. The hydrolytic conditions appear to have had no effect on the columns. Sample 2 varies slightly from those of 1, 3, and 4. We believe this arises from inadequately flushing the 1% H_2O solution from the chromatograph prior to analysis. Comparison of the data in Table 7 with the original data in Table 4 indicates abnormally high values for $\overline{M}_w$ and P. The broadening of the distribution appears to result primarily from tailing. Presently, we do not understand why the apparent $\overline{M}_w$ has increased so substantially.

In Figure 11 we see that the data for PBS are also good. The curves are nearly identical except for a slight inflection on the high molecular weight side. The inflection is real since it becomes more pronounced at lower flow rates. The data in Table 8 are very good for samples 1 and 3. The lower values for sample 2 presumably arise because of incomplete flushing of the columns after the first 16 hr. hydrolysis. Unlike PMPS, the $\overline{M}_w$ and P values for PBS are fairly close to the absolute values. With the exception of the slight high molecular weight inflections PBS appears to be eluted normally.

These results indicate that reaction with hindered silylating agents can successfully eliminate interactions between hydrophilic labile column sites and PMPS. Such columns are routinely used today for the GPC analysis of PMPS and PBS.

IV. CONCLUSIONS

We have examined various factors which influence the elution behavior of poly(2-methylpentene-1 sulfone) on μ-Styragel columns.
Inclusion of acidic and hydrogen bonding solvents in the carrier
solvent markedly influences the elution behavior perhaps by specific
protonation or solvation of the sulfone linkage in this polymer. We
surmise that ethanol in the $CHCl_3$ and absorbed water in the cyclohexanone might be responsible for the unusual effects observed by Ouano
and coworkers for poly(bicyclo[2.2.1]heptene-2 sulfone). In the absence of such interactions, PMPS is sorbed by the μ-Sytragel. Evidence for this behavior is the successful reproducible elution on
silica Zorbax columns for which sorption is precluded. Analysis of
5 PMPS samples in ethyl acetate using the Zorbax columns has revealed
the four highest molecular weight samples to have bimodal molecular
weight distributions. A log-log plot of [η] versus $\overline{M}_w$ indicates that
meaningful $\overline{M}_w$ values are obtained only for $\overline{M}_w$ values less than
1.0×10^6 g/mole.

   The bimodal nature of the chromatograms is caused by the interaction of hydrophilic sites on the support with the sulfone polymer.
Further silylation of the columns or inclusion of propylene glycol
in the eluting solvent eliminates the bimodality. It also changes
the elution behavior of PMPS. The effects are particularly disas-
trous when diols are included in the eluting solvent. These methods
do not permanently change the columns nor provide a useful analytical
technique for analysis of PMPS and PBS. We have found that reaction
of the columns with *tris*-trimethylsiloxysilyl esters does permanently
protect the labile sites on the columns. Successful reproducible
elutions of PMPS and PBS in ethyl acetate have been obtained. However, PMPS is observed to have too high a molecular weight.

ACKNOWLEDGMENT

We thank E. M. Doerries for PMPS samples and the [η] data, M. J.
Bowden for helpful discussions, and representatives of E. I. Dupont
for helpful suggestions regarding site protection.

REFERENCES

1. M. J. Bowden and L. F. Thompson, J. Appl. Polym. Sci., 17, 321
    (1973).

2. M. J. Bowden and L. F. Thompson, Polym. Eng. and Sci., 17, 269
    (1977).

3. L. F. Thompson, M. J. Bowden, E. M. Doerries, and S. R.
    Fahrenholtz, paper presented 8th Int. Conf. on Electron and Ion
    Beam Sci. and Technol., Electrochem. Soc. Meeting, Seattle,
    Wash., May 21-26, 1978.

4. M. J. Bowden and L. F. Thompson, Polym. Eng. and Sci., 14, 525
    (1974).

5. A. C. Ouano, E. Gipstein, W. Kaye, and B. Dawson, Macromol., 8,
    558 (1975).

6. A. C. Ouano, J. Coll. and Inter. Sci., 63, 275 (1978).

7. M. Y. Hellman and G. E. Johnson (in press).

THE APPLICATION OF GEL-PERMEATION CHROMATOGRAPHY
TO POLYOLEFIN PRODUCT PROBLEMS

Lowell Westerman

Plastics Technology Division
Exxon Chemical Company
Baytown, Texas

I. INTRODUCTION

Polyolefin resins such as polypropylene (PP), high density
polyethylene (HDPE) and low density polyethylene (LDPE), are employed in a wide variety of end use applications. These include the
major areas of film, fibers, and moldings. Specific applications
within each of these major end use areas demand resins with considerably different properties. For example, in the film area, desired
properties may include high clarity, high stiffness and high tensile
strength for one application, while good tear strength, controlled
shrinkage and low permeability may be required for another application. Often a number of different fabrication methods are employed
to produce the end product. The fabrication conditions combine with
the molecular structure of the resin and its rheological properties
to give a particular set of properties in the finished article. In
addition to physical property requirements, the resin customer wants
a material which is uniform, processes well on his equipment, and is
interchangeable with a resin from some other supplier. These demand-
ing requirements make product or grade development activities a very
challenging and often frustrating endeavour.

In years past, much of this type of work was conducted on an empirical trial-and-error-basis. Today, gel-permeation chromatography (GPC) as well as other improvements in polymer characterization techniques have greatly reduced the amount of trial-and-error associated with product development and has facilitated the effective solution of product and process problems.

This paper reviews the type of information which may be obtained by GPC, describes how this information may be applied to the solution of polyolefin problems, and presents examples of the application of GPC together with other information to the solution of practical problems encountered in actual practice.

II.   INFORMATION DEVELOPED BY GPC

Table 1 shows the various types of information which may be developed by GPC for linear and for long-chain branched polymers. It is commonly accepted that GPC separation takes place by size exclusion based on the "hydrodynamic volume," $[\eta]M$, of the polymer molecules in solution [1]. As a result, the primary information obtained is a "hydrodynamic volume" distribution. With the proper treatment of the data, this may be transformed into a molecular weight distribution (MWD) and any of the various average molecular weights such as number-average, $\overline{M}_n$, weight-average, $\overline{M}_w$, or Z-average, $\overline{M}_z$, may be computed. For linear polymers, the intrinsic viscosity may be calculated from the viscosity-average molecular weight if information is available for the Mark-Houwink constant and exponent for the solvent and temperature in question for the particular polymer of interest. Of course molecular weight dispersity parameters, such as $\overline{M}_w/\overline{M}_n$ and $\overline{M}_z/\overline{M}_w$, may be obtained. If GPC data is combined with dilute solution viscosity measurements it is possible to obtain molecular weights for long-chain branched polymers such as LDPE, and to determine the degree of long-chain branching as well as various molecular size averages such as $(Mg)_w$ and size dispersity ratios [2-5]. These size averages, involving the product of

TABLE 1

Information Developed by GPC

	Linear Polymers	Branched Polymers[a]
Hydrodynamic volume distribution	X	X
MW distribution	X	X
MW averages, $\overline{M}_n$, $\overline{M}_w$, $\overline{M}_z$, etc.	X	X
Intrinsic viscosity	X	--
MW dispersity ratios, $\overline{M}_w/\overline{M}_n$, etc.	X	X
Degree of long-chain branching	--	X
Long-chain branching distribution	--	X[b]
Molecular size averages, $(\overline{Mg})_n$, $(\overline{Mg})_w$, etc.	--	X
Molecular size dispersity ratios, $(\overline{Mg})_w/(\overline{Mg})_n$	--	X

[a]Requires a measured intrinsic viscosity on the sample.
[b]Requires use of low laser light scattering detector.

molecular weight and the branching function, g, are proportional to
the mean-square radius of gyration of the polymer molecules in solution.   Information may also be obtained about long-chain branching
distribution within a sample, if molecular weights are determined
directly on the eluant from the GPC by use of a low-angle laser
light scattering detector [6].

Obviously, quite a bit of information relative to the molecular
structure of a polyolefin resin may be developed by GPC together with
auxiliary measurements such as intrinsic viscosity or light scattering. This information by itself is of little value. It must be related to polymer polymerization and to polymer property information
before its real potential may be realized. Figure 1 shows the
interrelationships which exist in a grade development situation.
There are polymerization and post-polymerization variables to con-
tend with. These variables influence the molecular structure of the
resin obtained. We must be aware of the influence of these variables if we expect to alter or to control the molecular structure of
the resin. Molecular structure influences physical properties of
the polymer such as melt viscosity, melt elasticity, crystallinity,
melting behavior, and morphology. These physical properties, in
turn, interact with the processing or fabrication operation to yield
a product with a specific set of mechanical or optical properties.

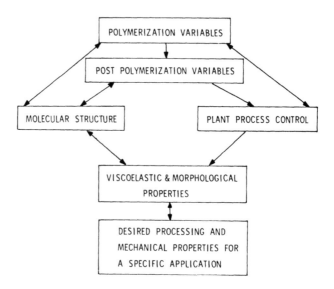

FIGURE 1.   Interrelationships existing in a polyolefin product
development situation.

When a desired product has been identified in terms of the rheological, mechanical and optical properties desired, we must be able to
translate this information back through molecular structure to the
polymerization variables in order to make the resin. We must also
define process control tests which will assure that the plant is
maintained in control to produce the desired product.

GPC information fits intimately into this total picture. Its
effective use demands a high degree of interaction between people
generating GPC data and those who use this information. Without
effective communications and use of appropriate auxiliary information, the full potential GPC cannot be realized and in some cases
GPC data may be misleading. The two examples below illustrate pitfalls to be avoided in the application of GPC data.

Consider two polypropylene reactor resins PP-2 and PP-9 whose
GPC curves are shown in Figure 2. The catalyst researcher submitt-
ing the samples wants to know if the molecular weight distribution
of the two samples are different. The answer is obvious just from
an inspection of the GPC curves. Yes, the molecular weight

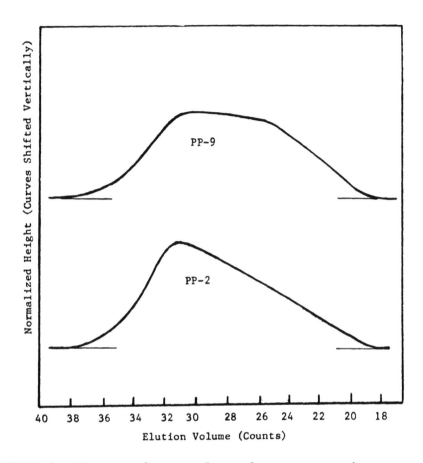

FIGURE 2.   GPC curves for two polypropylene reactor resins.

distributions of PP-2 and PP-9 are very different. This answer,
however, is misleading. If a series of polypropylene reactor resins
of different $\overline{M}_w$ or melt flow rate (MFR) are examined, the set of GPC
curves shown in Figure 3 is obtained. A definite trend in the shape
of the GPC curves may be noted in progressing from low molecular
weight (high MFR) to high molecular weight (low MFR). The change in
molecular weight distribution with the molecular weight level of the
resin is characteristic of the polymerization process. The two
samples PP-2 and PP-9 are shown again in Figure 3, and fit nicely
into this pattern. Thus, a more complete answer may be given based
on this background. Yes, the molecular weight distributions of

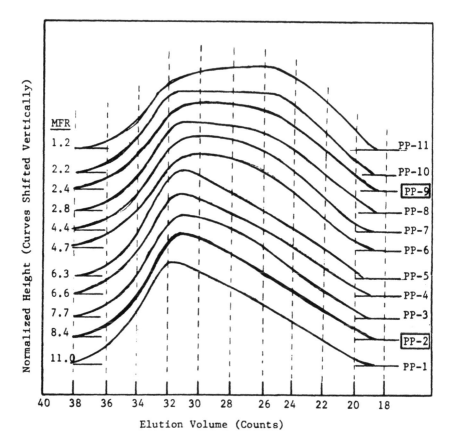

FIGURE 3. GPC curves for a series of polypropylene reactor resins of varying $\overline{M}_w$ or MFR.

these two samples are different, so are their $\overline{M}_w$'s; however, the difference is not unusual, it is what should be expected when the difference in MFR or $\overline{M}_w$ of these two reactor products are taken into consideration. The incomplete answer to the question could have led the catalyst researcher to believe that he has made a catalyst which was capable of giving an unusual molecular weight distribution when, in fact, he had not. This example illustrates that considerable care needs to be exercised in drawing conclusions from isolated comparisons of molecular weight distribution information.

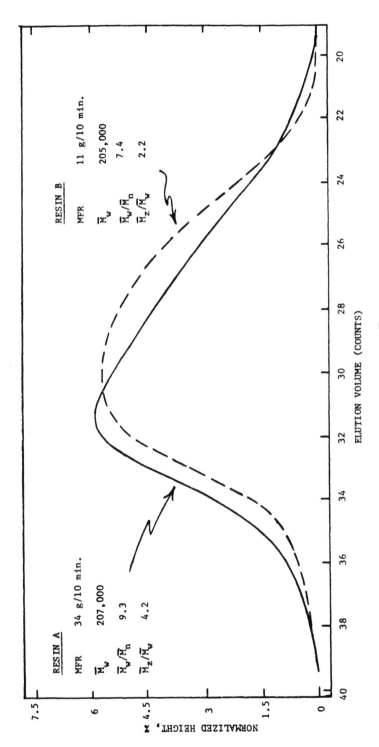

FIGURE 4. GPC curves for polypropylene resins having the same $\overline{M}_w$ but different MWD.

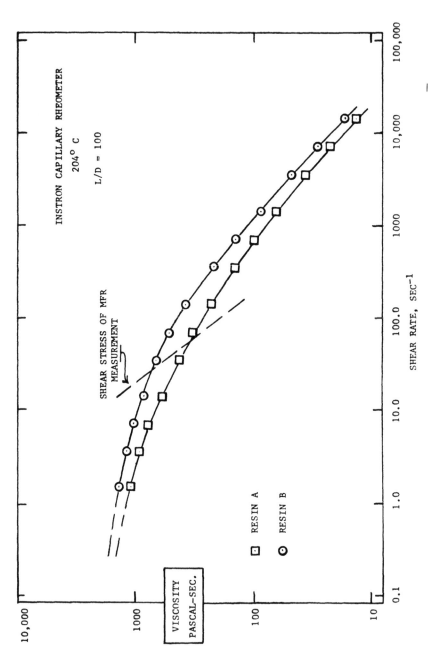

FIGURE 5. Melt viscosity versus shear rate for polypropylene resins having the same $\overline{M}_w$ but different MWD. ⊡ Resin A, MFR = 34 g/10 min.; ⊙ Resin B; MFR = 11 g/10 min.

Another case involves a product development engineer who was overheard saying, "GPC can't tell the difference between a 10 MFR and a 35 MFR polypropylene resin. They tell me that the $\overline{M}_w$'s are the same." Figure 4 shows the GPC curves and data for resins which might fit this case. The $\overline{M}_w$ values are the same: 205,000 and 207,000. However, $\overline{M}_w/\overline{M}_n$ values are somewhat different, 7.4 and 9.3, and a more marked difference is observed in $\overline{M}_z/\overline{M}_w$, 2.2 and 4.2. These differences in molecular weight distribution are reflected in the melt rheological behavior of these two resins as shown in Figure 5. Melt viscosity is plotted versus shear rate on a log-log scale. At very low shear rates the viscosity of resin A and B converge toward the same low shear limiting viscosity as expected from the equality of $\overline{M}_w$ for the two resins. The broader molecular weight distribution resin, A, shows a greater decrease in viscosity with increasing shear rate than does the narrower distribution resin B. The shear stress at which the MFR measurement is made is shown by the dashed line on the plot. At this shear stress resin A displays an appreciably lower viscosity than does the narrower MWD resin B in agreement with the difference in MFR noted.

The two examples given above are intended to emphasize the need for effective communications between the GPC practitioner and the user of GPC information and the need for background knowledge about the relationship between GPC information and the important polymer properties such as melt rheological behavior.

III.   APPLICATION OF GPC TO POLYOLEFIN PROBLEMS: EXAMPLES

A.   Polypropylene Melt Elasticity: A Plant Problem

In addition to the melt viscosity behavior of polyolefin resins, melt elasticity is an important melt characteristic for a number of end use applications such as fibers, film, and shape extrusion. Melt elasticity is known to be highly dependent upon the nature of the MWD [7]. As a consequence, GPC has been an important tool in developing resins with controlled melt elasticity.

Various melt rheological measurements may be made to obtain
information on melt elasticity. One very common technique involves
a simple measurement of the extent of swelling (swell) of an extru-
date from a capillary viscometer die under a fixed set of conditions.
When a polymer melt is pushed through a cylindrical capillary die of
diameter d_o it is found to swell to a larger diameter, d_e, upon
exiting from the die. The ratio, d_e/d_o or sometimes $(d_e/d_o)^2$, is
used as a measure of the melt elasticity of the polymer melt.

Most polypropylene manufacturers adjust the melt elasticity or
swell of their resins by means of a post polymerization operation
which essentially involves controlled molecular weight breakdown of
reactor resin of high molecular weight and broad molecular weight
distribution. The effect of chain scission of carbon-carbon bonds
on the molecular weight and molecular weight distribution has been
described by Davis, Tobias, and Peterli [8] and by Kotliar [9].
Chain scission reduces the molecular weight and alters the MWD of
the resin. For broad MWD resins such as polypropylene the MWD is
narrowed by random chain scission. For example, Figure 6 shows GPC
curves for a typical polypropylene reactor resin and for the product
obtained from this resin by controlled chain scission. The as-
polymerized resin may have a $\overline{M}_w$ of 500,000 and $\overline{M}_w/\overline{M}_n$ of 10-12. This
resin may be reduced to $\overline{M}_w$ of 100,000 and $\overline{M}_w/\overline{M}_n$ of about 4 by controlled scission.

For certain application areas, polypropylene grades may be produced which carry a melt elasticity specification. These grades
require rigid post polymerization process control. The problem described below demonstrates the application of GPC, background information developed by combined GPC and melt elasticity studies, and
some detective work to the solution of a plant problem with one of
these polypropylene grades having a critical swell specification.
The problem is illustrated in Figure 7 where melt elasticity as
measured by swell, $(d_e/d_o)^2$ is plotted versus melt flow rate for
products produced by controlled molecular scission in the laboratory
and at the plant. It is found that the plant produced products

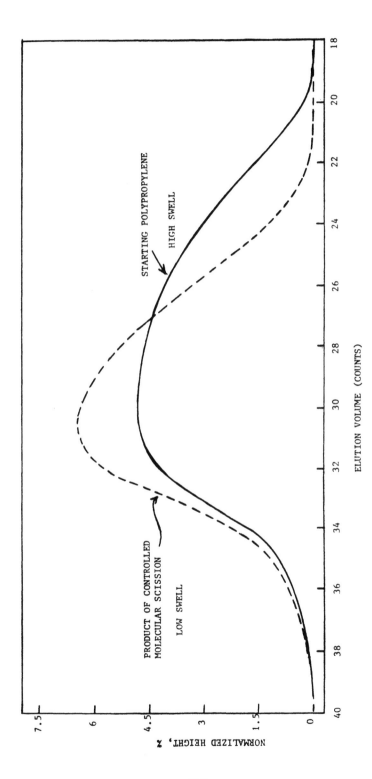

FIGURE 6. GPC curves for a typical broad MWD, high MW polypropylene resin (————), and the product obtained by controlled molecular scission (—————).

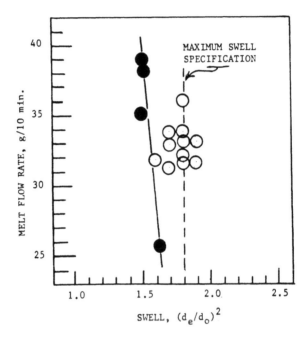

FIGURE 7. MFR vs. swell, $(d_e/d_o)^2$, for products from controlled
molecular scission. ⊙ Plant produced products; ○ laboratory prepared products.

consistently have larger swell values than laboratory products and
are dangerously close to failing the maximum swell specification
previously established for the product.

Typical GPC curves for a laboratory and plant produced resin
are shown in Figure 8. The difference between these curves is small,
and indeed challenges the limit of the precision of GPC to differentiate between the MWD of the two resins. Faced with this very small
difference in the GPC curves for these two resins it is tempting to
say that MWD is not a factor in this problem; however, if we take
the curves at their face value, the plant produced resin appears to
contain slightly more very high molecular weight species than the
product produced in the laboratory. Replication of these GPC curves
allows one to show that the two MWD's are different. The question
remains as to whether or not this small difference in MWD is

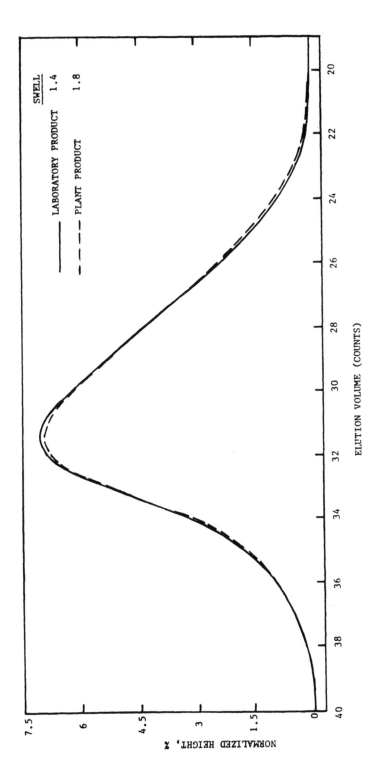

FIGURE 8. GPC curves for products from controlled molecular scission. Plant produced product (– – – – –); laboratory prepared product (————).

269

sufficiently large to account for the swell differences noted between
these samples.

We have found that swell values obtained on polypropylene
samples covering a broad range in molecular weight and molecular
weight distribution can be correlated reasonably well by the use of
a GPC parameter which heavily weighs the very high molecular weight
molecules in the distribution. This correlation is shown in Figure
9. The parameter obtained from GPC data, $J_e(GPC) = 2/5(\overline{M}_z\overline{M}_{z+1}/\overline{M}_w)$,
is plotted against swell as a measurement of elasticity. The param-
eter $J_e(GPC)$ is related to the steady state shear compliance as given
by Ferry [7]. There is some scatter in the data; however, it is not
greater than the precision with which the molecular weight averages

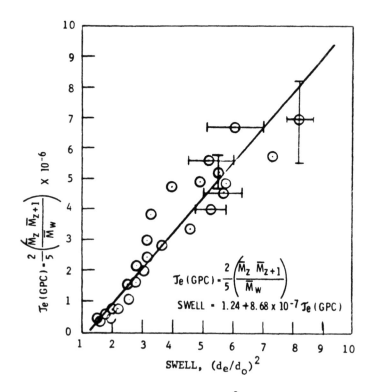

FIGURE 9. Correlation of swell, $(d_e/d_o)^2$ with $J_e(GPC)$ for polypropylene resins differing widely in $\overline{M}_w$ and MWD.

entering into J_e(GPC) can be determined. The major factor to be
considered is that this correlation as expressed by equation 1
below, allows us to predict the sensitivity of swell to a change in
the very high molecular weight region of the molecular weight
distribution.

$$\text{Swell} = 1.24 + 8.68 \times 10^{-7} \, J_e(\text{GPC}) \qquad\qquad (1)$$

It may be concluded that the small difference in MWD noted between
the plant and laboratory produced resins is sufficient to cause the
difference in swell noted.

The fact that the difference in MWD between the laboratory and
plant produced samples is sufficient to cause the difference in the
swell of the products does not help the plant solve the problem unless we can determine where this difference in MWD originates. Some
detective work uncovered the information that the plant resins are
formulated with various additives to produce the commercial grades.
This is in contrast with the laboratory samples to which no additives
had been added. Further digging uncovered the fact that the additives were added by blending an additive masterbatch into the product
from controlled molecular scission. The masterbatch was made by mixing the additives into a high molecular weight, broad MWD resin. As
much as 20% of this masterbatch was added to produce some of the end
commercial products. It was postulated that the blending of the high
molecular weight, broad MWD polypropylene resin from the masterbatch
into the lower molecular weight, narrow MWD resin produced a broadening of the MWD sufficient to cause the increase in swell. The effect
of blending was simulated on the computer by taking the GPC curves
for the two components and adding these distributions in various proportions. The molecular parameters were calculated for each computer
synthesized blend. A series of GPC curves for these computer blends
are shown in Figure 10. This figure graphically illustrates the expected change in MWD in the very high molecular weight region due to
the plant practice of adding the broad MWD, high molecular weight
masterbatch to incorporate additives.

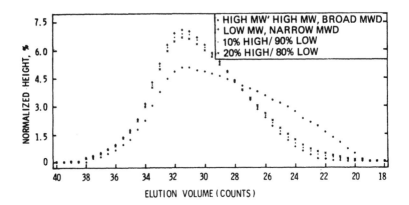

FIGURE 10. Experimental GPC curves for two blend components and computer synthesized blends. o High MW, Borad MWD blend component; + Low MW, narrow MWD blend component; ● Blend 10% High/90% Low; x Blend 20% High/80% Low.

Figure 11 further amplifies the way in which swell is expected to change with the amount of the high molecular weight component. Swell, in this figure, has been calculated from equation 1 which describes the correlation between J_e(GPC) and swell shown in Figure 9. It may be noted that swell increases rapidly with the incorporation of only a small amount of the high molecular weight component. Thus, it may be seen that only a few percent of the high molecular weight component is required to raise the swell by 0.3 units. These data dramatically demonstrated that the cause of the higher swell of plant produced products was the incorporation of a high molecular weight, broad MWD polypropylene masterbatch. The problem was resolved at the plant by incorporation of additives using a masterbatch made from controlled molecular scission polypropylene of similar molecular weight and MWD as the target product.

B.   Easier Processing Low Density Polyethylene:
     A Product Development Problem

Another example of the application of GPC to the solution of product problems involves a product development program concerned with finding out how to produce a LDPE resin with extrusion processability

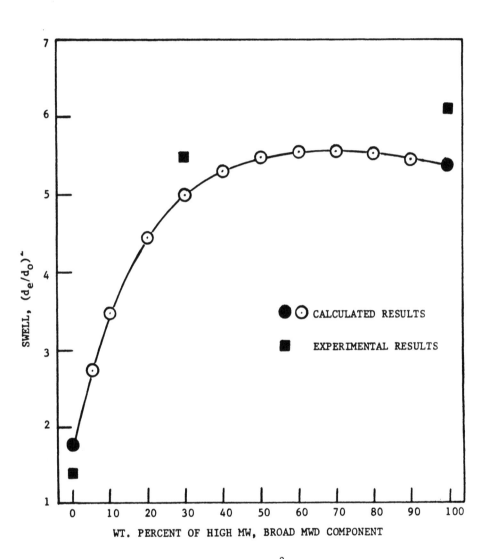

FIGURE 11. Variation of swell, $(d_e/d_o)^2$, with the concentration of high MW, broad MWD component in the blend. Calculated swell values from J_e(GPC) using equation 1.

equivalent to or better than the best of competition. The approach
to this problem involved two dependent parts:

1. Identification of the molecular features which control the shear
    thinning behavior of low density polyethylene melts, and

2. Identification of the polymerization process conditions which
    are necessary to alter the molecular structure in the manner de-
    sired.

The above two objectives were met through an integrated study of
LDPE molecular structure, melt rheology, and plant process conditions.

   Figure 12 gives a definition of the rheological terms, and
schematically illustrates the objective of this program. Melt
viscosity is shown plotted versus shear rate on a log log scale.
Two flow curves are shown with equivalent low shear limiting

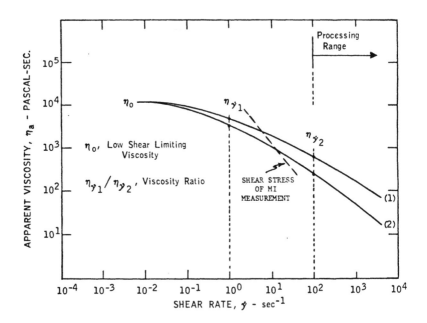

FIGURE 12. Definition of rheological parameters and schematic
representation of flow curves for low and high shear sensitive LDPE
resins. Curve (1), low shear sensitivity; curve (2), high shear
sensitivity.

viscosities, η_o. As the shear rate is increased, the viscosity of
both resins decreases due to the shear thinning behavior of the
melts. Shear sensitivity, the degree to which the viscosity changes
with shear rate, may be represented by the ratio of two viscosities
at a low and high shear rate, $\dot{\gamma}_1$ and $\dot{\gamma}_2$. The larger this ratio, the
greater will be the shear sensitivity of the resin. Thus, the resin
giving flow curve (2) has a higher shear sensitivity than the resin
giving flow curve (1).

For ease in extrusion processing a resin is desired with a low
viscosity at high shear rates, such as resin (2) compared with (1).
The objective, then, is to determine what molecular characteristics
influence the shear dependent viscosity behavior of LDPE resins, and
to relate these molecular characteristics back to the polymerization
process so that the desired changes in molecular structure may be
made.

For linear polymers it is well accepted that the low shear
limiting viscosity, η_o, depends upon $\overline{M}_w^{3.4}$ [10]. Shear dependent
viscosity is a function of not only $\overline{M}_w$, but also molecular weight
distribution. For randomly long-chain branched polymers the situation is somewhat different. It has been shown [4,11] that low shear
limiting viscosity of LDPE resins depends not on $\overline{M}_w$ but rather on
the weight-average molecular size, $(\overline{Mg})_w$, which is proportional to
the weight-average mean square radius of gyration $<S^2>_w$. The parameter $(\overline{Mg})_w$ can be obtained from GPC data combined with intrinsic
viscosity measurements [4,5]. The shear dependent viscosity should
be dependent upon $(\overline{Mg})_w$, MWD, and in addition upon the degree of
long-chain branching. Our study of shear dependent viscosity of
numerous LDPE resins has indicated that three molecular parameters
which can be obtained by combining GPC and intrinsic viscosity data
are important:

1. Weight-average molecular size, $(\overline{Mg})_w$.
2. Weight to number-average molecular weight ratio, $\overline{M}_w/\overline{M}_n$.
3. The number of long-chain branch points per weight-average mole-
    cular, $\overline{n}_w$.

The correlation between the viscosity ratio formed from viscosities measured at 1 sec^{-1} and 100 sec^{-1} can be expressed by the equation

$$\ln(\eta_1/\eta_{100}) = -22.2 + 2.3 \ln(\overline{Mg})_w - 0.12 \ln \overline{\eta}_w + 0.12 \ln \overline{M}_w/\overline{M}_n$$

(2)

with a correlation coefficient, $R = 0.965$.

Equation 2 indicates that $(\overline{Mg})_w$ is the most important parameter influencing the ratio of viscosities at two different shear rates, and that $\overline{\eta}_w$ and $\overline{M}_w/\overline{M}_n$ must also be considered in explaining differences in shear sensitivity of resins having the same $(\overline{Mg})_w$. The influence of increasing $\overline{\eta}_w$ at a fixed $\overline{M}_w/\overline{M}_n$ appears to be a reduction in shear sensitivity, while the influence of increasing $\overline{M}_w/\overline{M}_n$ at a fixed $\overline{\eta}_w$ increases shear sensitivity. An increase in $\overline{\eta}_w$ is naturally accompanied by an increase in $\overline{M}_w/\overline{M}_n$ in LDPE polymerizations. As a result, these two factors tend to compensate one another, and resins differing widely in $\overline{\eta}_w$ and $\overline{M}_w/\overline{M}_n$ can have virtually identical shear dependent viscosity behavior.

It is convenient to consider the effects of $\overline{\eta}_w$ and $\overline{M}_w/\overline{M}_n$ together in a combined form, the Branching-Molecular Weight Distribution Ratio (BDR) defined as:

$$BDR = (\overline{\eta}_w+1)/((\overline{M}_w/\overline{M}_n) - 1)$$

(3)

Using the BDR, the correlation equation corresponding to Equation 2 then becomes:

$$\ln(\eta_1/\eta_{100}) = -22.1 + 2.3 \ln(\overline{Mg})_w - 0.134 \ln BDR$$

(4)

Equation 4 indicates that resins of the same $(\overline{Mg})_w$ will have the same shear dependent viscosity behavior if their BDR's are the same. In other words, to alter shear dependent viscosity behavior, one needs to alter the ratio of long-chain branching to the breadth of the molecular weight distribution.

A large number of commercial LDPE resins have been examined in our laboratory. We have found it instructive to plot $(\overline{\eta}_w+1)$ versus $(\overline{M}_w/\overline{M}_n) - 1$ as shown in Figure 13 for a representative set of

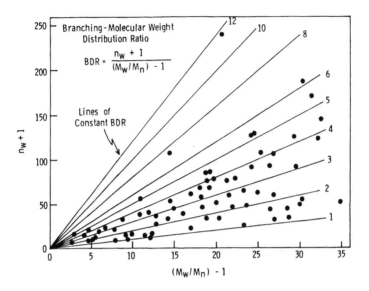

FIGURE 13. Relationship between long-chain branching and molecular weight distribution for commercial LDPE resins of constant $(\overline{Mg})_w$.

commercial resins. The lines on the plot represent constant values for the BDR. According to Equation 4, resins of constant $(\overline{Mg})_w$ will have the same shear dependent viscosity behavior along any of these lines of constant BDR. Easier processing resins, those with high shear sensitivities, fall along lines of low BDR while the less shear sensitive resins fall along lines of high BDR.

The information presented above, obtained from a study of the effect of molecular structure on the melt rheology of LDPE resins, when combined with information about plant polymerization conditions employed to make the various types of LDPE's, allowed us to determine the polymerization conditions needed in the plant to produce resins with altered BDR's and with improved extrusion processability. Plant runs were conducted which successfully produced resins with modified BDR. Figure 14 shows the output rate of an instrumented extruder divided by the torque on the extruder screw (a measure of extrusion processability) for two modified resins from plant test runs, together with the best of competition and standard LDPE resins. The high output/torque for any screw RPM is an indication of better

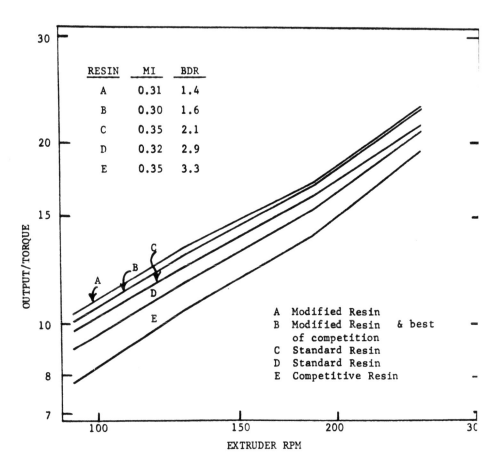

FIGURE 14.   Extruder output/torque on screw versus screw RPM for
extrusion of modified BDR and standard LDPE resins.

extrusion processability. It can be seen that the modified BDR
resins A and B show the highest output/torque for this series of
resins, and indeed resin B matches the best of competition and is
superior in processability to commercial resins C, D, and E.   The
objective of this program was met due to a total integrated program
in which GPC, melt rheology, and plant process data were employed.

IV.   CONCLUSIONS

A considerable amount of molecular structure information is
available for polyolefin resins from GPC together with auxiliary
measurements such as dilute solution viscosity measurements and low
angle laser light scattering. The effective use of this information
demands a high degree of interaction between people generating GPC
data and the people using this information. The examples of the
application of GPC data to polyolefin product problems illustrates
both pitfalls in attempting to use GPC information without proper
communication or with insufficient background information and total
integrated programs where GPC data is drawn together with other
information about the physical properties of the polyolefin and with
plant operating information. The latter approach, in the author's
experience, is the most effective way to apply the strength of GPC
to solving polyolefin product problems.

V.   EXPERIMENTAL

All GPC data presented in this paper were obtained using a
modified Waters Associates Model 200 gel-permeation chromatograph
equipped with four 3/8 inch x 4 foot columns packed with CPG-10
controlled porosity glass obtained from Electro-Nucleonics. Nominal
pore sizes were 100, 500, 1000, and 2000 Å. Column temperature was
maintained at 140°C. The elution solvent was 1,2,4-trichlorobenzene
stabilized with 0.02 wt. % 2,6-di-t-butyl-t-methyl phenol (BHT).
Samples for injection were prepared in fresh solvent from the instrument at a concentration of 0.25 wt. %. An injection time of 1 min
with a flow rate of 1 ml/min was employed to give a column loading
of 0.0025 g. Detection of polymer concentration in the eluant was
accomplished by monitoring the carbon-hydrogen stretching absorption
band at 3.5 microns using a Wilks Scientific Corp. Miran IA infrared
detector with an IR cell path length of 2 mm. No corrections were

applied to the data for band spreading. Details of long-chain
branching measurements for LDPE samples have been reported elsewhere
[4,5].

REFERENCES

1. H. Benoit, A. Grubisic, P. Rempp, D. Decker, and J. Zilliox, J.
    Chem. Phys., 63, 1507 (1966).

2. E. E. Drott, Polymer Preprints, May 1967, p. 67.

3. E. E. Drott, and R. A. Mendelson, J. Poly. Sci., A2, 8, 1361
    (1970).

4. C. I. Chung, J. C. Clark, and L. Westerman, in Advances in
    Polymer Science and Engineering (K. D. Pae, D. R. Morrow, and
    V. Chen, eds.), Plenum Press, New York, 1972, p. 249.

5. L. Westerman, and J. C. Clark, J. Poly. Sci., Poly. Phys. Ed.,
    11, 559 (1973).

6. A. C. Ouano and W. Kaye, J. Poly. Sci., Chem. Ed., 12, 1151
    (1974).

7. J. D. Ferry, Viscoelastic Properties of Polymers, 2nd ed., John
    Wiley, New York, 1970, p. 252.

8. T. E. Davis, R. L. Tobias, and E. B. Peterli, J. Poly. Sci.,
    56, 485 (1962).

9. A. M. Kotliar, J. Poly. Sci., A2, 1057 (1964).

10. G. C. Berry and T. G. Fox, Adv. Polymer Sci., 5, 261 (1968).

11. R. A. Mendelson, W. A. Bowles, and F. L. Finger, J. Poly. Sci.,
    A2, 8, 105 (1970).

DETERMINATION OF POLYETHYLENE MELT INDEX FROM GPC DATA

William A. Dark

Waters Associates, Inc.
Milford, Massachusetts

I. INTRODUCTION

It is well known, primarily from practical experience, that the
rheological properties of polymeric materials are strongly influenced
by molecular weight, molecular weight distribution, long chain branching and, possibly, reactor configuration. The theoretical development
of these relationships are far from complete, although they have been
addressed by several experimenters [1-7].

In this work, a relationship between melt flow rate (melt index)
and molecular weight of linear and branched polyethylene was explored.

A relationship relating isothermal melt viscosity and the molecular weight of a polymer was first expressed by Fox and Flory [9].

$$\log = 3.4 \log M_w + A$$

where A is an empirical constant which is characteristic of the polymer, but independent of molecular weight.

II. EXPERIMENTAL

A series of branched (ICI autoclave process) and linear (Phillips
process) polyethylene were evaluated.

Melt index was determined in accordance with ASTM D-1238 [8], density in accordance with ASTM D-1505-.

Molecular weight distribution and molecular weights were determined by gel permeation chromatographic techniques. The experimental conditions of the Model 150C GPC are given in Table 1.

The GPC system used Styragel columns, and their calibration curve is shown on Figure 1.

III.   RESULTS

Using standard GPC techniques, the molecular weight distributions and average molecular weights of the samples were determined. The molecular weight averages along with melt index, density, and crystallinity are reported in Table 2. Crystallinity values were determined by x-ray techniques.

Varying relationships between melt index and weight average molecular weight and viscosity average molecular weight were examined.

The relationship that yielded a reasonable correlation is shown in Figure 2. A relationship between log (MI) versus weight average molecular weight. This is a variation of the Fox-Flory relationship noted above. But, theirs is a relationship using intrinsic viscosity,

TABLE 1

GPC Conditions, M150C

Columns	μStyragel
	10^6Å, 10^5Å, 10^3Å
Solvent	1,2,4-Trichlorobenzene
Flow rate	1.0 ml/min
Column temp.	139°C
injector temp.	139°C
Run time	38 min
Injection volume	375
Sample conc.	0.1 wt. %
Sensitivity	128
Scale factor	25
Syringe motor speed	0.8 ml/min
Bottle spin	Normal
Filter time	1.0 min

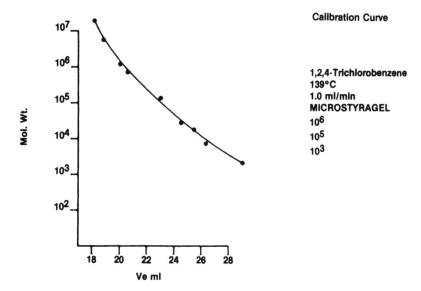

FIGURE 1. Calibration curve: polethylene molecular weight vs. retention volume. GPC M-150-C 1,2,4 trichlorobenzene at 1.0 ml/min at 139°C across a bank of three (10^6Å, 10^5Å, 10^3Å) μSTYRAGEL columns.

where here melt index is being used. In the relationship between melt index and weight average molecular weight, two families of curves were found: one for branched and the other for the linear polyethylenes.

These relationships yield:

Branched

$$\log MI = (0.454 \times 10^{-5})\bar{M}w - 0.12713 \qquad (1)$$

Linear

$$-(\log MI) = (0.6723 \times 10^{-5})\bar{M}w - 0.49359 \qquad (2)$$

The data scatter of these two simple relationships are indicative of their limited predictabilities. These relationships do not account for:

Long chain branching

Nonrandom termination

Reactor boundary conditions

TABLE 2

MI	Density	% Crys.	$\overline{M}n$	$\overline{M}w$	$\overline{M}_{vis}$	(η)
Linear						
0.81	0.960	85.5	12,300	151,100	101,300	1.7703
0.73	0.960	85.5	15,400	271,900	168,400	2.5590
0.30	0.960	85.5	16,900	179,800	114,100	1.9299
0.24	0.9555	79.5	20,100	165,600	112,900	1.9152
0.30	0.9505	72.5	15,900	132,300	91,500	1.6442
0.20	0.9385	56.5	14,900	171,900	117,700	1.8736
0.30	0.9505	72.5	22,400	157,900	114,400	1.9337
Branched						
1.32	0.9245	35.0	19,000	95,800	72,200	1.3852
6.95	0.9270	27.5	15,900	149,800	104,600	1.8121
11.61	0.9270	27.5	17,100	218,300	140,100	2.2394
0.80	0.9245	35.0	17,000	92,400	69,500	1.3470
2.07	0.9298	45.5	17,700	60,500	50,700	1.0721
6.56	0.9182	40.0	21,700	262,400	163,400	
7.22	0.9182	40.0	18,900	249,900	155,400	
6.72	0.9182	40.0	17,000	229,500	146,400	
7.32	0.9812	40.0	18,000	206,000	135,900	
6.96	0.9182	40.0	17,700	213,600	139,700	
6.16	0.9182	40.0	17,400	208,100	136,200	
6.58	0.9182	40.0	17,500	214,400	139,700	
6.44	0.9182	40.0	16,500	225,600	143,900	
6.54	0.9182	40.0	14,300	225,500	142,200	
6.64	0.9182	40.0	17,400	218,400	141,800	
6.42	0.9182	40.0	16,400	232,200	148,700	
7.00	0.9182	40.0	17,500	207,600	135,800	
6.48	0.9182	40.0	14,800	204,500	132,200	
6.33	0.9182	40.0	14,800	202,600	130,400	
7.10	0.9182	40.0	12,400	225,200	136,700	
6.64	0.9182	40.0	9,300	203,300	121,900	

Even though the above expression represents the best fit of the $\overline{M}w$ of the branched polyethylene data points, it cannot be correct. This expression indicates that a higher $\overline{M}w$ yields a high melt index. This is not what experience indicates.

These branched polyethylenes cover a broad range of branching, reactor configuration, and telogen level. To eliminate a number of these variables, a relationship between melt index and average molecular weight of a family of ICI baffled autoclave polyethylenes, where no telogen is used, was explored.

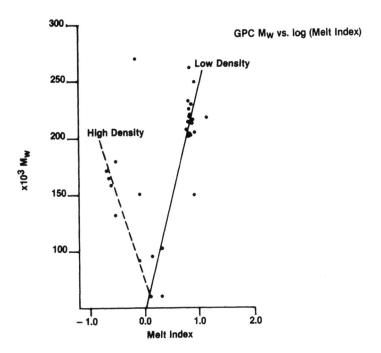

FIGURE 2.   Weight average molecular weight of linear and branched polyethylenes vs. log melt-flow viscosity.

Dealing with a single autoclave configuration where ΔT is held constant, then the contributions to multi-index of

Long chain branching

Telogen level

Reactor configuration

Molecular weight distribution

should be constant.   Only molecular weight averages should play the dominant role.

From the data two relationships were found:

$$MI = 14.287 - (3.595 \times 10^{-5})\overline{M_w} \tag{3}$$

$$\log(MI) = 0.7474 + (2.681 \times 10^{-4})\overline{M_z} \tag{4}$$

Calculation of melt index by each relationship versus determining melt index are shown in Table 3.

The mean variation between actual MI and that calculated from
$\overline{M}w$ is ±0.30 while that calculated from $\overline{M}z$ is ±0.14 g/10 min.

One of the polyethylene samples was analyzed nine times to
evaluate reproducibility.

$\overline{M}_n$ 16,200 ± 810

$\overline{M}_w$ 215,300 ± 8.360

$\overline{M}_z$ 2,279,000 ± 261,000

$\overline{M}_v$ 138,800 ± 3,990

Calculation of melt index from reproducibility data

Actual	From $\overline{M}w$	From $\overline{M}z$
6.44 g/10min	6.55 ± 0.30 g/10 min	6.44 ± 0.10 g/10 min

TABLE 3

	Calculated melt index	
Actual MI	From $\overline{M}w$	From $\overline{M}z$
6.56 g/10 min	4.85 g/10 min	7.30 g/10 min
7.22	5.30	7.18
6.72	6.04	6.60
7.32	6.88	6.30
6.96	6.61	6.36
6.16	6.81	6.33
6.58	6.58	6.38
6.44	6.18	6.55
6.54	6.18	6.60
6.64	6.44	6.43
6.42	5.94	6.43
7.00	6.82	6.35
6.48	6.94	6.34
6.33	7.00	6.39
7.10	6.19	7.16
6.64	6.98	6.95
AVERAGE VARIATION:	±0.30	±0.14

Note: The mean variation of melt index from actual values and calculated from GPC data and the variation from relative standard deviation of a single polyethylene are the same.

At a melt index of 6.0 g/10 min typical relative standard deviation of ASTM D-1238 is 0.15 g/10 min. Thus, with their family of ICI autoclave polyethylenes' Mz yields a useable relationship for melt index.

From similar autoclave conditions a series of branched polyethylenes were analyzed. The pressure drop across the autoclave was varied, while maintaining a constant temperature and initiator type for polyethylene formation. Thus long chain branching, reactor configuration, and nonrandom termination were held constant. The effect of changes in pressure were reflected in melt index and average molecular weight values. This relationship is shown in Figure 3.

The two NBS "standards" 1475, a linear polyethylene whole polymer and 1476, a branched polyethylene whole polymer were also evaluated.

NBS	Type	Mn	Mw	Mz
1475	Linear	14,200	103,500	6,901,600
1476	Branched	20,500	59,700	213,900

Using Eqs. 1 and 2 above, calculation for melt index yield:

	Determined	Calculated
1475	2.07 g/10 min	2.20 g/10 min
1476	1.19	1.39

Use of relationships 3 and 4 for NBS 1476 yield gross errors. It would be expected that NBS 1476 to very low in long chain branching, while the ICI autoclave polymer used to develop relationships 3 and 4 contains a much higher level of long chain branching.

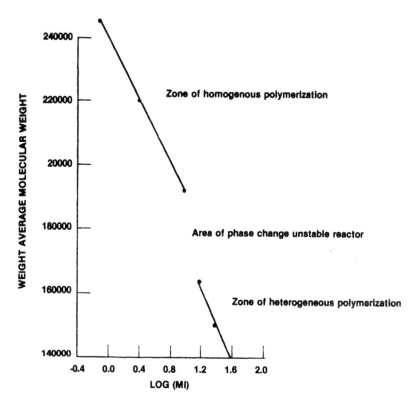

FIGURE 3.   Weight average molecular weight vs. log melt-flow
viscosity.   Samples of branched polyethylene from ICI autoclave
where pressure was varied to achieve differences in melt-flow
viscosity.

IV.   SUMMARY

Relationships between gel permeation chromatographic average molecular values and melt index of some polyethylenes has been explored.
Using $\overline{Mz}$ or $\overline{Mw}$ relationships have been developed that relate well
with ASTM determined melt index.   With branched polyethylene it has
been shown that consideration of long chain branching, nonrandom
termination, and reactor profile is necessary.   Linear polyethylene
appears to yield simpler relationships.

REFERENCES

1. Douglas P. Thomas, _Polymer Eng. Sci._, 11, 305–311 (1971).

2. G. Petraglia and A. Coeh, _Polymer Eng. Sci._, 10, 79–85 (1970).

3. A. N. Karasev, et al., Vysokomolekullannye Soedinenua, Ser A., 12, 1127–1137 (1970).

4. C. D. Han and C. A. Villamizar, _J. Appl. Polymer Sci._, 17, 2135 (1977).

5. A. F. Margolies, J.S.P.E., 27, 44–48 (1971).

6. B. H. Shah and R. Darby, _Polymer Eng. Sci._, 16, 579–584 (1976).

7. J. Thies and L. Schoenemann, _Advan. Chem. Series_, 109, 86–91 (1972).

8. American's Society for Testing and Materials, 1916 Race St., Philadelphia, Pa.

9. P. J. Flory, _Chem. Rev._, 39, 137 (1946).

AUTHOR INDEX

Author's names are followed by page numbers and, where appropriate, by reference numbers in parentheses.

SUBJECT INDEX

Acetophenone, 161
Acid strength, 243
Acrylamide, 183
Acylonitrile-butadiene-styrene,
    etc., 134
Albumin, 14
Alkyllithium, 38
Anionic polymerization, 36
Arrhenius parameters, 125

Bimodal columns, 107
Bisphenol A, 140, 159, 176, 183
Blake-Kozy-Carman equation, 225
Blending, 271
Block copolymers, 29
Bondagel, 107, 213
Bondapak, 170
Branch points, 275
Branched block copolymers, 39
Branching analysis, 47
Branching density, 30, 59
Branching distribution ratio,
    277
Branching function, 30
Branching parameter, 30, 59
B-stage, 159
Butadiene, 29

Chain cleavage, 238, 266, 269

Chain growth, 84
Chemisorption, 240
Coatings, 169
Coiled microcolumn, 127
Comonomer ratio, 34
Composite materials, 157
Compositional analysis, 41
Compressibility index, 23
Compression effect, 21
Compression ratio, 26
Concentration effects, 229
Concentration gradient, 10
Crystallinity, 282
Curable coatings, 182
Cyclic trimer, 101
Cyclohexane, 95

Deconvoluted peaks, 74
Degradation, 208, 221
Density, 282
Detector sensitivity correction,
    137
Dextran, 2, 14
Dicyanidiamide, 159
Dienes, 29
Diffusional non-equilibrium, 9
Diglycidyl ether, 159
dimethylformamide, 93
distribution coefficient, 4
dual detection, 42